SHEPHARD TECHNOLOGIES AND NEOCLASSICAL
PRODUCTION FUNCTIONS

Tilburg studies in econometrics
Vol. 2

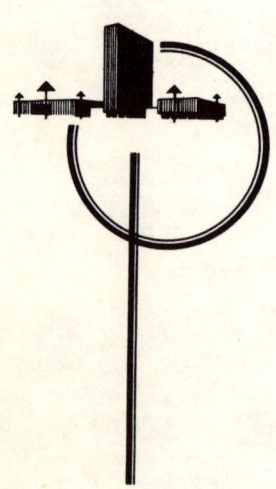

Shephard technologies and neoclassical production functions

TH. JUNIUS
University of Groningen,
the Netherlands

Martinus Nijhoff Social Sciences Division
Leiden 1977

ISBN 90 207 0727 2

© 1977 by H. E. Stenfert Kroese B.V. Leiden

No part of this book may be reproduced in any form by print, photoprint, microfilm or any other means, without written permission from the publisher

Printed in the Netherlands by Intercontinental Graphics Dordrecht

To the memory of my father

Contents

1. **Introduction / 1**

1.1. Preliminary remarks / 1
1.2. Purpose of the study / 3
1.3. Outline of the study / 4
Notes / 8

2. **Shephard technologies, neoclassical two-factor production functions and substitutability / 9**

2.1. Introduction / 9
2.2. Shephard technologies and neoclassical production functions / 9
2.3. Restricted functions and their properties / 19
2.4. Factor-substitution and its properties / 22
2.5. Reversed approaches / 25
2.6. The law of diminishing returns to factors / 30
Notes / 31

3. **Neoclassical production functions in economic dynamics / 34**

3.1. Introduction / 34
3.2. The neoclassical one-sector growth model / 35
3.3. Neutral technical change and neoclassical production functions / 41
Notes / 56

4. **The indirect specification of classes of neoclassical production functions / 58**

4.1. Introduction / 58
4.2. The indirect specification of the class of CES production functions / 59
4.3. The indirect specification of classes of VES production functions / 65
4.4. A class of VES production functions generating S-neoclassical production functions / 76
Notes / 82

5. **The direct specification of classes of neoclassical production functions / 84**

5.1. Introduction / 84
5.2. General remarks on the concept of approximation / 86
5.3. Introductory remarks on local approximation / 88
5.4. Local approximation of RC- and RCI-functions / 92
5.5. Local approximation of transforms / 103
5.6. Final remarks on local approximation / 112
Notes / 116

References / 119

Index of authors / 125

Index of subjects / 127

1. Introduction

1.1. PRELIMINARY REMARKS

1.1.1. The concept of a *production* technology is a basic notion in both theoretical and empirical economic investigations. In quantitative economic models it is possible to distinguish two methods in visualizing production technologies. First, in traditional neoclassical investigations concerning production technologies, the concept of a production *function* is used to express the technical relation between maximum output and the inputs of the various goods and services. Secondly, in modern activity-analysis approaches a production technology is defined by the use of abstract mathematical concepts from *set theory* and *topology*.

TAKAYAMA is of opinion that the second, activity-analysis, approach is more fundamental and more rigorous, (1). In many economic investigations, however, production functions are still being used. We believe that there are two important reasons for such "conservatism". First, taking a theoretical point of view, the body of general theorems which has been developed in neoclassical analyses is of a different scope than that developed in investigations bearing an activity-analysis character (2). Secondly, taking an empirical point of view, the application of neoclassical (differentiable) production functions often facilitates the use of econometric methods (3).

1.1.2. The seeming gap between the neoclassical production function approach and the activity-analysis approach in visualizing production technologies has been closed in the pioneering work of SHEPHARD (4). For the following reasons SHEPHARD's contributions are of fundamental importance:
1. He defines in an activity-analysis manner the properties of a (one output) production technology.
2. He analyzes the conditions under which such a production technology can be specified in an *equivalent* way by the use of the concept of a production function.

Or stated differently, SHEPHARD obtains a *one-to-one correspondence between his very general concept of a production technology and his very general concept of a production function*. For example, SHEPHARD's definition of a production function does away with continuity and differentiability assumptions.

1.1.3. SHEPHARD introduced his view of production technologies and production functions in 1953, but he does not believe the influences of this study have been very great (5). He reviewed and generalized this 1953 study in 1970. Although his 1970 contribution has had a greater influence on investigations regarding production technologies (6), SHEPHARD's views have never been applied consistently to the field of neoclassical thinking. E.g.:
1. PLASMANS tried to effectuate such an application, but disregarded SHEPHARD's *basic* assumption that the so-called *efficient subset* of each isoquant must be *bounded* (7).
2. KUIPERS explicitly started from the assumption that in his growth model, the efficient subset of an isoquant must be bounded, but he did not investigate whether his type of production function fits into a view of

production technologies (8).

1.2. PURPOSE OF THE STUDY

1.2.1. The purpose of this study is *to accomplish the direct and consistent application of SHEPHARD's framework to neoclassical thinking*. This is effected by:
1. The presentation of a "new" *definition* of the concept of a neoclassical *two-factor* production function in such a way that this concept fits into SHEPHARD's view.
2. The modification of a number of *theoretical* neoclassical studies by the application of this definition and by a reconsideration of the results obtained in these studies.
3. The investigation of functional forms on their possibility to generate production functions in accordance with the 'new' definition of a neoclassical two-factor production function. These functional forms have been introduced in economic literature and have been used in *empirical* investigations concerning two-factor production functions.

1.2.2. The SHEPHARD approach and the neoclassical approach are important from both a theoretical and an empirical standpoint. This means that the application of SHEPHARD's approach to the neoclassical approach may be fruitful. The main result of this study can be summarized as follows: if SHEPHARD's view of production technologies is accepted, then neoclassical thinking need not necessarily deviate from the lines initiated by SHEPHARD.

1.3. OUTLINE OF THE STUDY

1.3.1. For the two-factor case SHEPHARD's definition of a production technology is presented in chapter 2.
On the basis of this definition a neoclassical two-factor production function is defined in the following way; it is a *compound* function of a so-called *transform* and a so-called *core* function, where:
1. Both production factors are *essential*.
2. Both marginal factor productivities are strictly positive on a nonempty open convex *cone* in the strictly positive space of inputs.

The last assumption is a consequence of SHEPHARD's basic assumption that the efficient subset of an isoquant must be bounded.

The implied properties of the well-known neoclassical concepts of (i) the marginal rate of substitution and (ii) the elasticity of substitution are studied in chapter 2 with the aid of three functions, namely the *restricted* neoclassical production function, the *restricted* core function and the *restricted* core-intensity function.

It is shown that it is possible to modify and extend investigations of SATO and of FØRSUND, regarding the relationship between a *given* function of the elasticity of substitution and the implied class of neoclassical production functions.

Chapter 2 is ended with a counter-example concerning the Law of Diminishing Returns to Factors.

1.3.2. A number of *theoretical* investigations from the field of economic dynamics are modified in chapter 3 by applying the definition of a neoclassical production function.

The neoclassical one-sector growth model is the starting-point for the construction of a simple example. This example is used to investigate the implications of our way of defining a neoclassical production function regarding the conditions for balanced growth.

Next, a recent study by BECKMANN on neutral technical change is adapted. The discussion of this study, has a twofold character, namely:

1. It is argued that although BECKMANN defined in a traditional way the properties of a neoclassical production function, he did *not* investigate whether his proposed forms of neutral technical change generate functions which have the properties of a neoclassical production function.
2. The additional conditions on two forms of neutral technical change are presented, in such a way that these two forms yield production functions which fit into the definition of a neoclassical production function presented in chapter 2.

Regarding generalized BECKMANN neutral technical change it is shown that it can only be applied to a subclass of (aggregate) economies.

1.3.3. In *empirical* investigations functional forms are needed to quantify certain assumed relationships. As a starting-point for the quantification of productions technologies the concept of a neoclassical two-factor production function used to play and is still playing a prominent role. For such a case, it is often assumed that the relationship between maximum output and the inputs of the production factors is a neoclassical one. Mostly, however, without the explicit assumption of bounded efficient subsets.

In the specification of functional forms it is possible to distinguish two methods in the literature, *indirect* specification and *direct* specification. In chapter 4 a number of functional forms are discussed which have been specified by means of the indirect specification method. Our investigations seek an answer to the two questions:

1. Can these (indirectly specified) functional forms be used to generate neoclassical production functions in the sense of the definition of chapter 2? That is to say, do these functions fit into SHEPHARD's view?
2. Consider an arbitrary production function generated by the use of an indirectly specified functional form. Are the *largest possible* efficient subsets of this function bounded and do the boundaries of each subset only depend on the values of the parameters defining the corresponding functional form?

From an empirical viewpoint the answer to the last question is of basic importance. If this answer is in the affirmative, then *the indirectly specified functional form might be used to test the assumption of bounded largest possible efficient subsets*. If the answer is negative, this use is not possible.

Starting-point for the investigations in chapter 4 are two methods for obtaining indirectly the functional form of the well-known class of CES production functions. Based on these two methods an outline is given of a number of indirectly specified functional forms which generate VES production functions. It is shown that a class of VES production functions obtained independently by BRUNO and VAZQUEZ has the properties that the largest possible efficient subsets are bounded and that the boundaries of these subsets are determined by the values of the parameters of the functional form of the BRUNO/

VAZQUEZ-class. The CES-class and the other VES-classes do not have these properties.

1.3.4. In chapter 5 a number of functional form are investigated which have been defined in a direct way. It is argued that the direct choice for these functional forms can be based on the so-called *local approximation approach*. That is to say, these functional forms can be used for a local approximation to a given production function. If, however, it is assumed that the approximation itself must have the properties of a neoclassical production function, it is argued that the approximation must be decompounded into two independent steps:
1. A local approximation to either the given restricted function or the given restricted core-intensity function.
2. A local approximation to the given transform.

It is shown that functional forms introduced by KUIPERS, DIEWERT and DENNY respectively generate a function which:
1. Can perform a local second-order approximation to a given restricted core-function.
2. Has the properties of a restricted core function.
3. Has largest possible efficient subsets which are bounded.

The last property is, again, important for empirical applications, because these functional forms might be used to quantify the boundaries of the largest possible efficient subsets.

It is shown that the class of power functions (yielding homogeneity) and a functional form introduced by ZELLNER/REVANKAR generates a function which:
1. Performs a local approximation to a given transform.

2. Has the properties of a transform in the sense of the definition of a neoclassical production function.

To be able to investigate the properties of functional forms introduced by ZELLNER/REVANKAR and RINGSTAD the concepts of (i) the *scale-elasticity* function of a neoclassical production function and (ii) a general class of transforms introduced by FRISCH are discussed.

Thus, it is shown that the local approximation approach can be used in directly specifying functional forms which generate neoclassical production functions in the sense of the definition given in chapter 2.

NOTES

1. TAKAYAMA (1974), chapter 0, section C.
2. Despite his criticism TAKAYAMA uses neoclassical production functions when illustrating the application of mathematical methods from the theory of calculus of variations and optimal control, TAKAYAMA (1974), chapter 5, section D, and chapter 8.
3. For a survey see WALTERS (1963).
4. SHEPHARD (1953), (1967), (1970).
5. SHEPHARD (1970), Introduction.
6. See for example DIEWERT (1974b) and its comments. But in a recent textbook on microeconomics, however, no reference is made to the investigations of SHEPHARD, see SHONE (1975), chapters 5, 6 and 7.
7. PLASMANS (1975), chapter 1.
8. KUIPERS (1970), chapter 2.

2. Shephard technologies, neoclassical two-factor production functions and substitutability

2.1. INTRODUCTION

In section 2.2. of this chapter SHEPHARD's contribution to the definition of a production technology is discussed and a definition of a neoclassical production function is presented (1). In section 2.3. the definition of a neoclassical production function is leading to the definitions of a restricted neoclassical production function, a restricted core function and a restricted core-intensity function, followed by investigations concerning the implied properties of these last three functions. In section 2.4. the definitions of the marginal rate of substitution and the elasticity of substitution are presented and the properties of these two concepts are analyzed. In section 2.5. the relevant definition of a neoclassical production function leads to a modification and an extension of studies of SATO and of FØRSUND. This chapter concludes with a simple counter-example concerning the Law of Diminishing Returns to Factors.

2.2. SHEPHARD TECHNOLOGIES AND NEOCLASSICAL PRODUCTION FUNCTIONS

2.2.1. In this section an outline is given of SHEPHARD's view of a production technology. This outline concerns only the case of one output and two inputs (2). Subsequently SHEPHARD's view is applied to the concept of a neoclassical two-factor production function. Note that in SHEPHARD's 1970 study such an application was not effected. This means that in his 1970 study SHEPHARD did not investigate completely the relationships between his 1953 study and his 1970 study.

2.2.2. According to SHEPHARD a technology 'consists of a family of conceivable and feasible engineering arrangements, not restricted necessarily to particular realizations found in practice and possibly encompassing historical changes in the application of the technology' (3). If the set of all input-output combinations is known, then full information is also given about the production possibilities of a production technology. By means of concepts from the field of set-theory and topology, SHEPHARD mathematically defines a production technology in an activity-analysis way (4).

It is assumed that within a certain fixed period some good or service is obtainable as the *only* output by using the inputs of *two* distinct production factors (5). Let U denote the output rate where (6):

$$U \in \mathbb{R}_+ \tag{2.1}$$

and let X denote the row-vector of input rates of the two production factors where:

$$X \in \mathbb{R}_+^2. \tag{2.2}$$

DEFINITION 2.1: For each $U \in \mathbb{R}_+$ the *production input set* L(U) is the set of all input vectors $X \in \mathbb{R}_+^2$ yielding at least the output rate U.

RUYS called the point-to-set mapping L a multifunction (7).

DEFINITION 2.2: For each $U \in \mathbb{R}_+$ the *efficient subset* E(U) of a production input set L(U) is:

$$E(U) := \{X \in L(U) \;/\; X' \leq X,\; X' \neq X \Rightarrow X' \notin L(U)\}. \tag{2.3}$$

From this definition it follows that an input vector X is efficient relative to a production input set L(U), if and only if less factor-input in any component does not yield the given output rate U (8).

SHEPHARD defines a production technology in terms of the mathematical properties which must be attributed to the production input sets L(U) and to the efficient subsets E(U), (9).

DEFINITION 2.3: A *production technology* is a family of production input sets satisfying:
A.1. $L(0) = \mathbb{R}_+^2$, $0 \notin L(U)$ for $U > 0$.
A.2. $X \in L(U)$, $X' \geq X \Rightarrow X' \in L(U)$.
A.3. If $0 \neq X \leq 0$ and for some scalar $h^o > 0$ and some $U^o > 0$: $h^o X \in L(U^o)$, then the half-ray $\{hX \, / \, h > 0\}$ intersects $L(U)$ for all $U \in \mathbb{R}_+$.
A.4. For $0 \leq U \leq U'$ is $L(U') \subset L(U)$.
A.5. For each $U' > 0$: $\bigcap_{0 \leq U \leq U'} L(U) = L(U')$.
A.6. $\bigcap_{U \in \mathbb{R}_+} L(U) = \emptyset$.
A.7. For each $U \in \mathbb{R}_+$: the set $L(U)$ is closed.
A.8. For each $U \in \mathbb{R}_+$: the set $L(U)$ is convex.
A.9. For each $U \in \mathbb{R}_+$: the set $E(U)$ is bounded.

SHEPHARD is of opinion that properties A.1., ..., A.9. are to be taken 'as valid for any technology'. In this study each technology which exhibits these properties is called an S- (=SHEPHARD) technology. This to emphasize SHEPHARD's contribution to the definition of a production technology and to stress that different views of production technologies are possible (10). According to SHEPHARD properties A.1., ..., A.9 can be interpreted as follows, (11):
1. A nonnegative vector of input rates yields at least

the zero output rate and a zero input vector yields
no positive output rate, A.1.
2. Disposability of inputs, or excess of capacity implies
merely that the excess be foregone, A.2.
3. An input vector yielding at least an output rate $U_2 \geqq U_1$, also yields at least U_1, A.4.
4. An infinite output rate cannot be obtained efficiently from a finite input vector, A.6.
5. Any output rate can be obtained by scalar-multiplication of some vector of input rates, though maybe inefficiently, A.3.
6. The technology is time divisibly-operable, A.8.
7. No output rate is obtained efficiently from an unbounded vector of input rates, A.9.

Properties A.5. and A.7 are introduced to guarantee the existence of a production function and in order to be able to define the production isoquant.

In particular property A.9. is crucial in the difference between the traditional neoclassical approach and the approach of SHEPHARD. SHEPHARD argues (12) 'we shall adjoin this property (...) in order to have production level sets (in this study these are called production input sets) which are technologically realistic' and he emphasizes that (13) 'it is unreasonable to expect the input of one factor of production to be substitutable efficiently for another in unbounded amount'. This *basic* assumption excludes a number of traditional production functions like the well-known CES functions. PLASMANS for example based his definition of a neoclassical production function on the definition of an S-technology, but he disregarded SHEPHARD's assumption of bounded efficient subsets (14). This means that PLASMANS' analysis cannot be considered as a consistent application of the views of SHEPHARD to neoclassical thinking, (15).

DEFINITION 2.4: For each $U \in \mathbb{R}_+$ the *production isoquant* $I(U)$ of a production input set $L(U)$ is (16):

$$I(U) := \{X \in L(U) \, / \, \lambda X \notin L(U), \, 0 \leq \lambda < 1\}. \qquad (2.4)$$

It is clear from defs. 2.1, 2.2 and 2.4 that:

$$E(U) \subset I(U) \subset L(U), \, U \in \mathbb{R}_+. \qquad (2.5)$$

In figures 2.1a and 2.1b below we present two illustrations given by SHEPHARD (17). In figure 2.1a the two production factors can only be used efficiently in a fixed proportion, a so-called WALRAS-LEONTIEF technology. In figure 2.1b a technology is illustrated with four alternatives of mixing production factors. Both figures illustrate the fact that the family of production input sets $L(U)$ are describing the input-constrained technical possibilities.

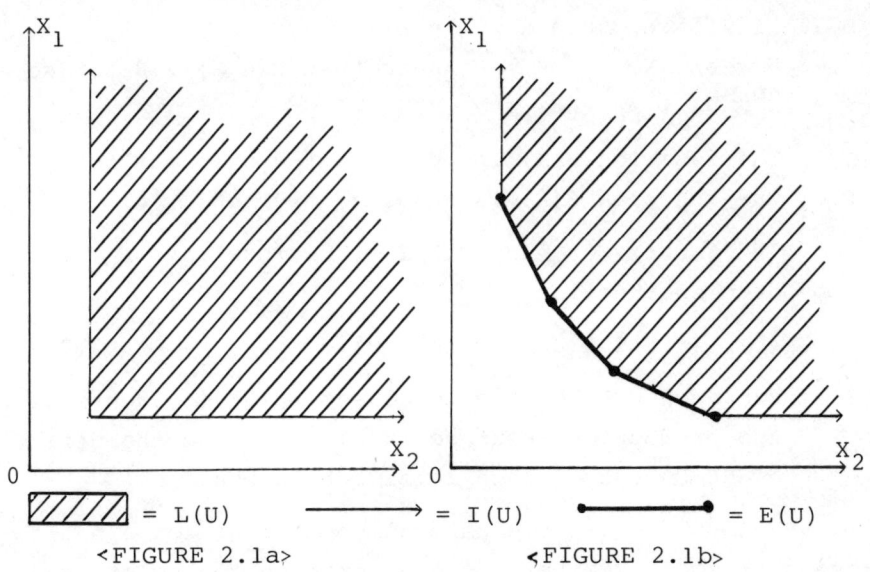

<FIGURE 2.1a> <FIGURE 2.1b>

2.2.3. SHEPHARD considers the production function as a mathematical concept defined on the production input sets of a production technology (18).

DEFINITION 2.5: The function H from $\mathbb{R}_+^2$ to $\mathbb{R}_+$ is a *production function* of a production technology, if and only if:

$$Y = H(X) = \max \{U \in \mathbb{R}_+ \,/\, X \in L(U)\}. \qquad (2.6)$$

This definition gives to the production function H its traditional meaning: Y is the *largest* output rate obtainable with the vector X of input rates.

SHEPHARD shows that, if the production input sets of a production technology have properties A.1., ..., A.9., the production function H has the following properties (19):

B.1. $H(0) = 0$.
B.2. For each $X \in \mathbb{R}_+^2$, the value $H(X)$ is finite.
B.3. If $X' \geqq X$, then $H(X') \geqq H(X)$.
B.4. For any X, $0 \neq X \geqq 0$, such that $H(h^o X) > 0$ for some $h^o > 0$: $\lim_{h \to \infty} H(hX) = \infty$.
B.5. The function $H(X)$ is upper semi-continuous on $\mathbb{R}_+^2$.
B.6. The function $H(X)$ is quasi-concave on $\mathbb{R}_+^2$.

DEFINITION 2.6: For each $U \in \mathbb{R}_+$ the *level set* $L_H(U)$ of a production function H is:

$$L_H(U) := \{X \in L(U) \,/\, H(X) \geqq U\}. \qquad (2.7)$$

B.7. For each $U \in \mathbb{R}_+$ the level set $L_H(U)$ is identical to the production input set $L(U)$ and hence the efficient subset of each level set is bounded.

SHEPHARD deduces the general conditions for a one-to-one correspondence between a production technology and a production function. He states: 'the structure may be defined either in terms of a suitable production function H or by a family of production input sets L(U), since one is uniquely determined from the other' (20), our symbols. This result may be called SHEPHARD's *equivalence* theorem. This theorem shows that there is no gap between the activity-analysis approach and the production function approach in the theory of production.

The properties of a production function H of an S-production technology are quite general. This means that one may ask whether it is possible to fit the traditional neoclassical approach within SHEPHARD's framework. In order to effect such a modification of the concept of a neoclassical production function we shall first introduce two basic concepts. These concepts were also used by SHEPHARD in his theory of production (21).

DEFINITION 2.7: Let H be a production function of an S-production technology. The function:

$$W(H(X)) = W(Y) \qquad (2.8)$$

is called a *transform* of the function H, if and only if $W(0) = 0$ and W is an upper semi-continuous and nondecreasing function of the variable Y.

SHEPHARD proves (22) that a transform W is a production function of some S-production technology, if
$$\lim_{Y \to \infty} W(Y) = \infty .$$

DEFINITION 2.8: Let H be a production function of an S-production technology. The function H is *homothetic*, if and only if H is of the form:

$$H(X) = W(\hat{H}(X)) = W(Z), \qquad (2.9)$$

where:
1. The function $\hat{H}$ represents an S-production technology; the function $\hat{H}$ may be called *core* or *kernel* function (23), the variable Z may be interpreted as a scalar

measure (index, aggregate) of the input vector (24).
2. The function $\hat{H}$ is homogeneous of degree one (25).
3. The function W is a transform with $\lim_{Z \to \infty} W(Z) = \infty$.

In this study we shall use the following notations:
1. The superscript ˆ over a *capital* letter indicates a function which is homogeneous of degree one with respect to the variables X_1 and X_2.
2. The superscript ˆ over a *small* letter a function which is homogeneous of degree zero with respect to X_1 and X_2.

SHEPHARD proves that (26), if the function H is homothetic, then $\hat{H}$ is super-additive (27), continuous and concave on $\mathbb{R}^2_+$, and for each $U \in \mathbb{R}_+$ there exists a scalar $a \in \mathbb{R}_+$ such that the isoquant I(U) can be obtained from the unit isoquant I(1) by radial expansion from the origin with the scalar a.

2.2.4. It is now possible to present a "new" definition of the concept of a neoclassical two-factor production function which fits into SHEPHARD's framework; for an alternative definition see PLASMANS (28).

DEFINITION 2.9: A function H is called a *neoclassical* production function of an (two-factor) S-production technology, if and only if the function H is homothetic with:
1. The transform W is twice continuously differentiable on $\mathbb{R}^2_{++}$, (29), $W' > 0$ and $\lim_{Z \to \infty} W(Z) = \infty$.
2. There exists a (convex) cone:

$$\bar{D} := \{X \in \mathbb{R}^2_{++} \; / \; a \leq X_1/X_2 \leq b\}, \; 0 < a < b < \infty, \quad (2.10)$$

such that:
2.1. The core function $\hat{H}$ is continuous on $\bar{D}$.
2.2. On the interior of $\bar{D}$ (D := interior $\bar{D}$) the function $\hat{H}$ is twice continuously differentiable.
2.3. On the cone D:

$\hat{H}_1' > 0, \hat{H}_2' > 0,$ (2.11a)

$\hat{H}_{11}''(\hat{H}_2')^2 - 2\hat{H}_{12}''\hat{H}_1'\hat{H}_2' + \hat{H}_{22}''(\hat{H}_1')^2 < 0.$ (2.11b)

2.4. For each $X \in \mathbb{R}_{++}^2$ with $X_1 > bX_2$ (see the area above the half-ray $X_1 = bX_2$ in figure 2.2 below):

$\hat{H}(X) = \hat{H}(bX_2, X_2).$ (2.11c)

2.5. For each $X \in \mathbb{R}_{++}^2$ with $X_1 < aX_2$ (see the area below the half-ray $X_1 = aX_2$ in figure (2.2):

$\hat{H}(X) = \hat{H}(aX_2, X_2).$ (2.11d)

2.6. If either $X_1 = 0$ or $X_2 = 0$, then:

$\hat{H}(X) = 0.$ (2.11e)

2.2.5. It is clear that in def. 2.9 the function H is a *compound* function of the functions W and $\hat{H}$. It can be easily proved that (i) in def. 2.9 the function $\hat{H}$ is unique up to a constant and that (ii) ineqs. (2.11b) and (2.11c) hold for the function H itself. We shall now consider the problem whether def. 2.9 generates a class of production functions representing S-production technologies (for an illustration of a unit isoquant of the core function $\hat{H}$ see figure 2.2).

If $X \in \bar{D}$ we find along the isoquant of $\hat{H}$:

$d\hat{H} = \hat{H}_1' dX_1 + \hat{H}_2' dX_2 = 0,$ (2.12)

or, from ineqs. (2.11a):

$dX_1/dX_2 = -\hat{H}_2'/\hat{H}_1' < 0.$ (2.13)

From eq. (2.12) we obtain:

$d^2\hat{H} = \hat{H}_{11}'' dX_1 dX_1 + 2\hat{H}_{12}'' dX_1 dX_2 + \hat{H}_{22}'' dX_2 dX_2 = 0,$ (2.14)

or, from ineqs. (2.11b) and eq. (2.12):

$d^2X_1/dX_2^2 = \dfrac{-\hat{H}_{11}''(\hat{H}_2')^2 + 2\hat{H}_{12}''\hat{H}_1'\hat{H}_2' - \hat{H}_{22}''(\hat{H}_1')^2}{(\hat{H}_1')^3} > 0.$ (2.15)

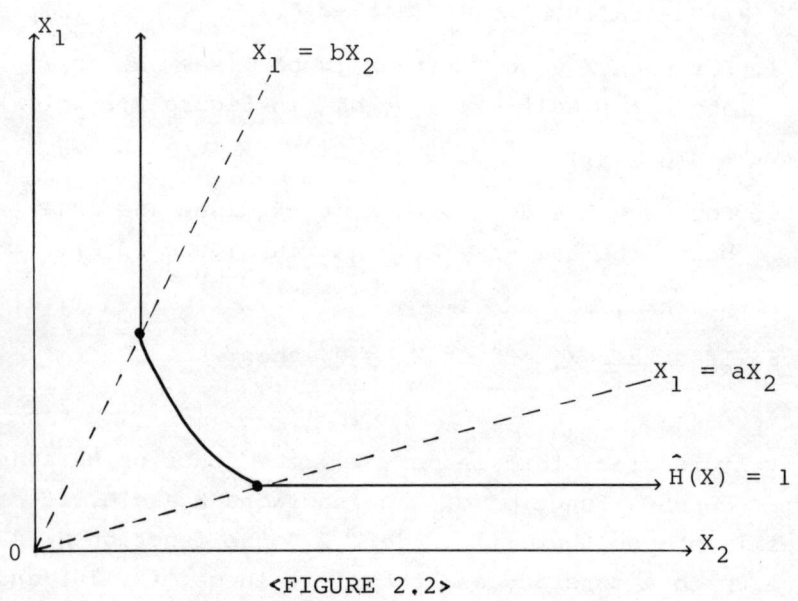

<FIGURE 2.2>

From ineqs. (2.13) and (2.15) we conclude that along each efficient subset E(Y) in D, the variable X_1 is a strictly decreasing, strictly convex function of the variable X_2, (30).

Eqs. (2.11c) and (2.11d) define each isoquant outside the cone $\bar{D}$ in a WALRAS-LEONTIEF way.

Eq. (2.11e) states that the two production factors are assumed to be *essential*. This means that no strict positive output rate can be obtained if at least one of the input rates is zero (31). For the case of two production factors we are of opinion that from a technological point of view this assumption is the only realistic one.

From (2.10) it follows that $\bar{D} \subset \mathbb{R}^2_{++}$, this implies that each efficient subset E(Y) is bounded. Since the function $\hat{H}$ itself represents an S-production technology, the compound function $H(X) = W(\hat{H}(X))$ represents an S-production technology. This means that def. 2.9 actually *links our neoclassical two-factor approach with the more general approach of SHEPHARD*. This statement results from the implied property that each level set $L_H(Y)$ of a neoclassical production function has properties A.1., ..., A.9. of a production input set L(Y) (see def. 2.3).

In this study a function H is called S- (=SHEPHARD) neoclassical (32), if and only if H is neoclassical in the sense of def. 2.9. If a function is S-neoclassical we shall sometimes call the function H homothetic in a neoclassical sense.

Observe that ineq. (2.11c) can be written as a restriction on the bordered HESSIAN, namely:

$$- \begin{vmatrix} 0 & \hat{H}'_1 & \hat{H}'_2 \\ \hat{H}'_1 & \hat{H}''_{11} & \hat{H}''_{12} \\ \hat{H}'_2 & H''_{12} & H''_{22} \end{vmatrix} > 0. \qquad (2.16)$$

This restriction is more general than (i) the restriction (33):

$$H''_{ii} < 0, \; i = 1,2, \qquad (2.17)$$

or (ii) the restriction that the HESSIAN is negative definite (34). See also the discussion in section 2.6.

2.3. RESTRICTED FUNCTIONS AND THEIR PROPERTIES

2.3.1. In this section we shall introduce three so-called restricted functions. The restricted neoclassical production function is used in order to be able to neglect that part of $\mathbb{R}^2_+$ over which nonefficient production will occur (in figure 2.2 the area outside $\bar{D}$). The restricted core-intensity function is introduced:
1. To represent the properties of a given restricted neoclassical production function in a compact way.

2. To investigate the possibilities for factor-substitution in technologies which are represented by S-neoclassical production functions.

One of the purposes of this study is to apply the definition of an S-neoclassical production function to a number of economic subjects. It is shown in the forthcoming chapters that these two functions will be useful in obtaining several results in a consistent way. The restricted core function can be regarded as a link between a given restricted neoclassical production function and its related core-intensity function.

2.3.2.

DEFINITION 2.10: Let H be an S-neoclassical production function. The function F is the *restricted neoclassical production function (= RP-function)* of H, if and only if F is the restriction of H on the cone $\bar{D}$, i.e.:

$$F(X) := \begin{cases} H(X) \text{ for } X \in \bar{D}, \\ \text{not defined for } X \notin \bar{D} \end{cases} \quad (2.18)$$

From this definition we conclude that each isoquant $I_F(Y)$ of an RP-function F equals $E_H(Y)$. This means that each RP-function disregards nonefficient production.

DEFINITION 2.11: Let H be an S-neoclassical production function. The function $\hat{F}$ is the *restricted core function (= RC-function)* of H, if and only if $\hat{F}$ is the restriction of $\hat{H}$ on the cone $\bar{D}$.

From defs. 2.9, 2.10 and 2.11 it follows that:
1. Each function $\hat{F}$ is homothetic, i.e. $F(X) = W(\hat{F}(X))$.
2. On the open cone D:

$$\hat{F}'_1 > 0, \ \hat{F}'_2 > 0, \quad (2.19a)$$

$$\hat{F}''_{11}(\hat{F}'_2)^2 - 2\hat{F}''_{12}\hat{F}'_1\hat{F}'_2 + \hat{F}''_{22}(\hat{F}'_1)^2 < 0. \quad (2.19b)$$

DEFINITION 2.12: For each $X \in \bar{D}$ the *input-ratio* x is:

$$x := X_1/X_2, \quad a \leq x \leq b. \tag{2.20}$$

DEFINITION 2.13: Let H be an S-neoclassical production function. The function $\hat{f}$ is the *restricted core-intensity function* (= *RCI-function*) of H, if and only if:

$$\hat{f}(x) = \hat{F}(x,1), \text{ domain } \hat{f} = [a,b]. \tag{2.21}$$

2.3.3. In this subsection it will be shown that the RCI-function $\hat{f}$ has properties which are of a very simple analytical character. First, we conclude from eq. (2.21) that $\hat{f}$ is continuous on [a,b] and that the function $\hat{f}(X_1/X_2)$ is homogeneous of degree zero with respect to X_1 and X_2. Secondly, eq. (2.21) yields:

$$\hat{F}'_1 = \hat{f}', \tag{2.22a}$$

$$\hat{F}'_2 = \hat{f} - x\hat{f}'. \tag{2.22b}$$

From these inequalities we obtain that on (a,b) the function $\hat{f}'$ exists and is continuous and:

$$\hat{f} > 0, \tag{2.23a}$$

$$0 < x\hat{f}'/\hat{f} < 1. \tag{2.23b}$$

The application of EULER's theorem to the function $\hat{F}$, $\hat{F}'_1$ and $\hat{F}'_2$ yields for each $X \in D$:

$$\hat{F}'_1 X_1 + \hat{F}'_2 X_2 = \hat{F}, \tag{2.24a}$$

$$\hat{F}''_{11} X_1 + \hat{F}''_{12} X_2 = 0, \tag{2.24b}$$

$$\hat{F}''_{12} X_1 + \hat{F}''_{22} X_2 = 0. \tag{2.24c}$$

From system (2.24) we find:

$$\hat{F}'_2 = (\hat{F} - X_1 \hat{F}'_1)/X_2, \tag{2.25a}$$

$$\hat{F}''_{12} = -X_1 \hat{F}''_{11}/X_2, \tag{2.25b}$$

$$\hat{F}''_{22} = X_1^2 \hat{F}''_{11}/X_2^2, \tag{2.25c}$$

while:

$$\hat{F} = X_2 \hat{f}, \qquad (2.26a)$$

$$\hat{F}'_1 = \hat{f}', \qquad (2.26b)$$

$$\hat{F}''_{11} = \hat{f}''/X_2. \qquad (2.26c)$$

Systems (2.24), (2.25) and (2.26) can be used for rewriting ineq. (2.19b), namely:

$$\hat{F}''_{11}(\hat{F}'_2)^2 - 2\hat{F}_{12}\hat{F}'_1\hat{F}'_2 + \hat{F}''_{22}(\hat{F}'_1)^2 = \hat{f}^2\hat{f}''/X_2^2. \qquad (2.27)$$

From eq. (2.27) we obtain a third property, on (a,b) the function $\hat{f}''$ exists and is continuous and:

$$\hat{f}'' < 0. \qquad (2.28)$$

The investigations of this subsection are summarized in the next theorem.

THEOREM 2.1: If the function H is a given S-neoclassical production function, then the RCI-function $\hat{f}$ has the following properties:
1. The function $\hat{f}$ is homogeneous of degree zero.
2. Domain $\hat{f}$ = [a,b] $\subset \mathbb{R}^2_{++}$.
3. The function $\hat{f}$ is twice continuously differentiable on (a,b).
4. On (a,b): $\hat{f} > 0$, $0 < x\hat{f}'/\hat{f} < 1$, $\hat{f}'' < 0$.

2.4. FACTOR-SUBSTITUTION AND ITS PROPERTIES

2.4.1. In the neoclassical theory of production the notion of *factor-substitution* is well-known and standard (35). Two functions have been introduced to measure main aspects of this notion, the marginal rate of substitution and the elasticity of substitution. In this section these two functions are defined and analyzed in terms of a given S-neoclassical production function H. It must be noticed that incorrect definitions of measures for factor-substitution may lead to incorrect results.

2.4.2.

DEFINITION 2.14: Let H be an S-neoclassical production function. The function $\hat{s}$ from D to $\mathbb{R}_{++}$ is the function of the *marginal rate of substitution* (= *MRS-function*) of H, if and only if:

$$\hat{s}(X) = F_1'/F_2' . \tag{2.29}$$

From ineqs. (2.19a) it follows that the function $\hat{s}$ exists on D. Along each isoquant $I_F(Y)$ we obtain:

$$\hat{s}(X) = -(dX_2/dX_1)_{F(X)=\text{const.}} , \tag{2.30}$$

which means that the function $-\hat{s}$ can be interpreted as, if the use of the first factor is increased by one unit, $-\hat{s}$ measures the number of units by which the use of the second factor must be reduced, to keep the level of output at a constant level. Or, the function $-\hat{s}$ is a measure of the slope at the isoquant through $X \in D$. Furthermore, we find from system (2.22) that:

$$\hat{s}(x) = \hat{f}'/(\hat{f} - x\hat{f}'), \text{ domain } \hat{s}(x) = (a,b). \tag{2.31}$$

From eq. (2.31) it follows that the MRS-function can be written as a function of the input-ratio only, or, the function $\hat{s}$ is homogeneous of degree zero. So along each half-ray in D, starting in the origin, the value of the function $\hat{s}(X) = \hat{s}(X_1/X_2)$ does not change.

From the properties of an RCI-function $\hat{f}$ it follows that the function $\hat{s}(x)$ is continuous and strictly positive on (a,b). From:

$$\hat{s}'(x) = \hat{f}\hat{f}''/(\hat{f} - x\hat{f}')^2 \tag{2.32}$$

we conclude that the function $\hat{s}'(x)$ exists, and is continuous and strictly negative on (a,b). The next theorem summarizes the properties of the function $\hat{s}$.

THEOREM 2.2: If the function H is a given S-neoclassical

production function, then the MRS-function $\hat{s}$ has the following properties:
1. The function $\hat{s}$ is homogeneous of degree zero.
2. Domain $\hat{s}(x) = (a,b)$.
3. The function $\hat{s}$ is continuously differentiable on (a,b).
4. On (a,b): $\hat{s} > 0$, $\hat{s}' < 0$.

2.4.3.

DEFINITION 2.15: Let H be an S-neoclassical production function. The function $\hat{\sigma}$ from D to $\mathbb{R}_{++}$ is the function of the *elasticity of substitution (= ES-function)* of H, if and only if:

$$\hat{\sigma}(X) := \frac{-F_1'F_2'(X_1F_1' + X_2F_2')}{X_1X_2(F_{11}''(F_2')^2 - 2F_{12}''F_1'F_2' + F_{22}''(F_1')^2)}. \quad (2.33)$$

From ineq. (2.19b) it follows that the function $\hat{\sigma}$ exists on D. Along each isoquant $I_F(Y)$ the function $\hat{\sigma}$ may be written as (36):

$$\hat{\sigma}(X) := -(\frac{dX_1/dX_2}{X_1/X_2} \frac{F_1'/F_2'}{d(F_1'/F_2')}) F(X)=\text{const.} \quad (2.34)$$

i.e.:

$$\hat{\sigma}(x) = -(dx/x)(\hat{s}/d\hat{s}). \quad (2.35)$$

The function $\hat{\sigma}$ may be interpreted as a measure for 'the ease with which the varying factor may be substituted for the other' (37). Or, the function $\hat{\sigma}$ is a normalized measure of the change in the function $\hat{s}$, see eq. (2.35)

The insertion of eqs. (2.31) and (2.32) into (2.35) yields after rearranging:

$$\hat{\sigma}(x) = -\hat{f}'(\hat{f} - x\hat{f}')/x\hat{f}''\hat{f}, \text{ domain } \hat{\sigma}(x) = (a,b). \quad (2.36)$$

Like the function $\hat{s}$, the function σ is homogeneous of degree zero, its value does not change along each half-ray in D starting in the origin. From eq. (2.36) and theorem 2.1 it follows that the function $\hat{\sigma}$ is continuous and strictly positive on (a,b). The investigations in this subsection lead to the next theorem.

THEOREM 2.3: If the function H is a given S-neoclassical production function, then the ES-function $\hat{\sigma}$ has the following properties:
1. The function $\hat{\sigma}(X)$ is homogeneous of degree zero.
2. Domain $\hat{\sigma}(x) = (a,b)$.
3. The function $\hat{\sigma}$ is continuous on (a,b).
4. The function $\hat{\sigma}$ is strictly positive on (a,b).

2.4.4. It is of fundamental importance to note that from the properties of an S-neoclassical production function H it follows that:
1. Both functions $\hat{s}$ and $\hat{\sigma}$ are invariant with respect to monotone transformations applied to a given S-neoclassical production function, see eqs. (2.31) and (2.36).
2. In the approach of this study the domains of $\hat{s}$ and $\hat{\sigma}$ are proper subsets in $\mathbb{R}^2_{++}$, i.e. the cone D. In a neoclassical approach it is incorrect to suggest that these two functions exist for negative values for at least one of the marginal factor productivities (38). The acceptance of SHEPHARD's view of a production technology leads to the impossibility of sketching isoquants which on certain parts may have positive slopes (39). This statement results from theorems 2.2 and 2.3.

2.5. REVERSED APPROACHES

2.5.1. In section 2.4 we investigated the properties of the functions $\hat{s}$ and $\hat{\sigma}$. A reversed approach has been proposed by SATO and by FØRSUND. These authors (40) started from a *given* ES-function $\hat{\sigma}$ and obtained a general expression for the implied class of production functions.

In their reversed approach they, however, did not attribute certain properties to the function $\hat{\sigma}$ and they did not investigate whether the implied class of functions actually generates *production* functions. The investigations of SATO and of FØRSUND will be modified and extended as follows:
1. It will be assumed that the function $\hat{f}$ (subsect. 2.5.2), the function $\hat{s}$ (subsect. 2.5.3) and the function $\hat{\sigma}$ (subsect. 2.5.4) resp. is given.
2. It will be proved that in all three cases the implied class of functions generates S-neoclassical production functions.

2.5.2. In the first problem to discuss the function $\hat{f}$ is *given* with (i) domain $\hat{f}$ = [a,b] $\subset \mathbb{R}_{++}$, (ii) $\hat{f}$ twice continuously differentiable on (a,b) and (iii) on (a,b): $\hat{f} > 0$, $0 < x\hat{f}'/\hat{f} < 1$, $\hat{f}'' < 0$. That is to say, $\hat{f}$ has the properties of an RCI-function.

If we define the function $\hat{F}(X) := X_2\hat{f}(X_1/X_2)$, then what can be said about the properties of F, will this function be an RP-function? From the results in subsect. 2.3.3, it follows directly that $\hat{F}$ has the properties of an RP-function on an open cone D. But since $\hat{f}$ is continuous on a closed interval, it follows that $\hat{F}$ is continuous on the closure of the cone D, say $\bar{D}$. By using eqs. (2.11c),...,(2.11e) this RP-function $\hat{F}$ can be extended to an (unrestricted) core function $\hat{H}$. If also a transform W is given, a unique S-neoclassical production function H can be obtained. So there is *one* degree of freedom, the choice out of the class of transforms W, if a degree of freedom is regarded as the possible choice out of a class of permissible functions. The results of this subsection lead to the next theorem.

THEOREM 2.4: If the function $\hat{f}$ is given and has the properties of an RCI-function, then one can attribute a class of S-neoclassical production functions to $\hat{f}$, with the transform W as the degree of freedom.

2.5.3. Let now $\hat{s}$ be a given function with (i) domain $\hat{s}$ = $(a,b) \subset \mathbb{R}_{++}$, (ii) $\hat{s}$ continuously differentiable on (a,b) and (iii) on (a,b): $\hat{s} > 0$, $\hat{s}' < 0$. I.e., the function $\hat{s}$ has the properties of an MRS-function. For $\hat{f}' \neq 0 \neq \hat{f} - x\hat{f}'$ we can obtain the following differential equation from eq. (2.31):

$$d\hat{f}/\hat{f} = \hat{s}dx/(1 + x\hat{s}). \qquad (2.37)$$

We shall discuss the following problem: if $\hat{s}$ is given and $\hat{f}$ an arbitrary solution of eq. (2.37), what can be said about the properties of this solution $\hat{f}$, has $\hat{f}$ the properties of an RCI-function?

For each solution $\hat{f}$ we may write:

$$\ln \hat{f} = \int_{x_o}^{x} \hat{s}dv/(1 + v\hat{s}), \; x \in (a,b), \qquad (2.38)$$

for some scalar $x_o \in (a,b)$. It is more convenient to write eq. (2.38) in terms of indefinite integrals:

$$\hat{f} = \exp \int \hat{s}dx/(1 + x\hat{s}). \qquad (2.39)$$

Since eq. (2.39) contains an integration constant, our first conclusion is that to $\hat{s}$ we can attribute a class of functions $\hat{f}$. Each function $\hat{f}$ from this class is continuous and strictly positive on $[a,b]$. Take a function from this class and differentiate the right-hand side of eq. (2.39) with respect to x. This yields that $\hat{f}'$ exists and is continuous on (a,b) and:

$$\hat{f}'(x) = \hat{u}(x)\exp \hat{u}(x), \qquad (2.40)$$

where:

$$\hat{u}(x) := \hat{s}(x)/(1 + x\hat{s}(x)), \text{ domain } \hat{u} = (a,b). \qquad (2.41)$$

So, $\hat{f}'$ is strictly positive on (a,b) and:

$$0 < x\hat{f}'/\hat{f} < 1, \qquad (2.42)$$

on (a,b).

Differentiating the right-hand side of eq. (2.40) yields that $\hat{f}''$ exists and is continuous on (a,b) and that:

$$\hat{f}''(x) = (\hat{s}'/(1 + x\hat{s})^2)\exp \hat{u}(x) \qquad (2.43)$$

That is to say, on (a,b) the function $\hat{f}''$ is strictly negative, i.e. the function $\hat{f}$ has the properties of an RCI-function. Applying theorem 4.2 yields that to each solution $\hat{f}$ of eq. (2.37) we can attribute a class of S-neoclassical production functions.

The discussions in this subsection have furnished the proof of the next theorem.

THEOREM 2.5: If the function $\hat{s}$ is given and has the properties of an MRS-function, then one can attribute a class of S-neoclassical production functions to $\hat{s}$, with the transform W and a strictly positive integration constant as the degrees of freedom.

2.5.4. As a last step in this section, we shall now assume that $\hat{\sigma}$ is a given function with (i) domain $\hat{\sigma} = (a,b) \in \mathbb{R}_{++}$, (ii) $\hat{\sigma}$ continuous on (a,b) and (iii) $\hat{\sigma}$ strictly positive on (a,b), i.e. the function $\hat{\sigma}$ has the properties of an ES-function.

Eq. (2.35) leads after rearranging to the following first-order differential equation:

$$d\hat{s}/\hat{s} = -dx/x\hat{\sigma}(x), \qquad (2.44)$$

in which the function $\hat{s}$ is unknown. We shall now show that each solution $\hat{s}$ of eq. (2.44) has the properties of an MRS-function.

For each solution $\hat{s}$ we may write:

$$\ln \hat{s} = - \int_{x_1}^{x} dt/t\hat{\sigma}(t), \ x \in (a,b), \qquad (2.45)$$

or in terms of indefinite integrals:

$$\hat{s} = \exp - \int dx/x\hat{\sigma}(x). \tag{2.46}$$

From eq. (2.46) we conclude that we can attribute a class of functions $\hat{s}$ to the given function $\hat{\sigma}$, containing a strictly positive integration constant as the degrees of freedom. Each function $\hat{s}$ from this class is strictly positive and continuous on (a,b). From the differentiation of the right-hand side of eq. (2.46) we find that $\hat{s}'$ exists and is continuous on (a,b) and:

$$\hat{s}'(x) = -\hat{s}/x\hat{\sigma}(x), \tag{2.47}$$

i.e. strictly negative on (a,b). This means that each solution $\hat{s}$ of eq. (2.44) has the properties of an MRS-function. Using theorem 2.4 and theorem 2.5, and the discussions above, lead to the following theorem:

THEOREM 2.6: If the function $\hat{\sigma}$ is given and has the properties of and ES-function, then one can attribute a class of S-neoclassical production functions to $\hat{\sigma}$, with the transform W and two strictly positive integration constants as the degrees of freedom.

2.5.5. We believe that for empirical applications the results of this section are quite powerful. The functions $\hat{f}$, $\hat{s}$ and $\hat{\sigma}$ have simple analytical properties. As an example, the graph of some hypothetical function $\hat{\sigma}$ has been sketched in figure 2.3 below. Without further examination we conclude from theorem 2.6 that this particular ES-function generates S-neoclassical production functions. This means that it is not necessary to obtain explicit solutions for implied classes or to investigate the shape of an arbitrary isoquant. The results obtained in subsection 2.5.4. we elsewhere denoted as: a reconsideration of the SATO-FØRSUND theorem (41).

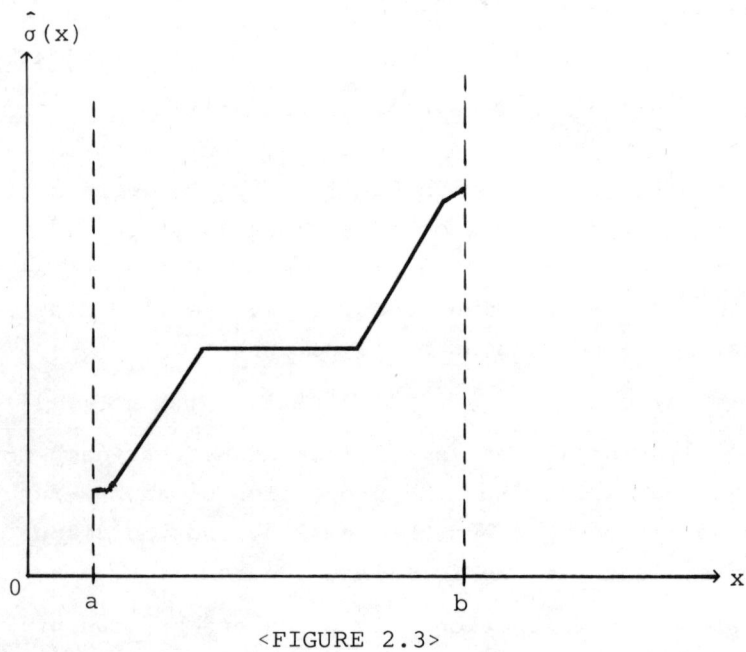

<FIGURE 2.3>

2.6. THE LAW OF DIMINISHING RETURNS TO FACTORS

2.6.1. In this section we shall show that the conditions on an RCI-function $\hat{f} > 0$, $0 < x\hat{f}'/\hat{f} < 1$, $\hat{f}'' < 0$, are *not* sufficient for the RP-function F to exhibit the so-called Law of Diminishing Returns to Factors, i.e. $F_i' > 0$, $F_{ii}'' < 0$, $i = 1,2$, on D (42).

It can be easily shown that, if F is homogeneous of a degree $h \leqq 1$ (nonincreasing returns to scale), this Law holds on D. A simple numerical example will now be given of an RP-function, which is homogeneous of a degree $h > 1$ (increasing returns to scale), buth which does not have the property of Diminishing Returns to Factors.

Let:

$$\hat{f}(x) = -(x - 1)(x - 9), \text{ domain } \hat{f} = [3,5]. \qquad (2.48)$$

It can be easily verified that on [3,5] (2.48) actually defines an RCI-function. Let the RP-function F be defined as follows:

$$F(X) = (X_1 - X_2)^2 (X_1 - 9X_2)^2, \qquad (2.49)$$

F homogeneous of degree 4. Take $X_2 = 1$, then for each $X_1 \in (3,5)$ we find that:

$$F_1'(X_1,1) > 0, \qquad (2.50)$$

but:

$$F_{11}''(X_1,1) \begin{cases} \geqq 0 \text{ for } 3 < X_1 \leqq (5 - \tfrac{4}{5}\sqrt{5}), & (2.51a) \\ < 0 \text{ for } (5 - \tfrac{4}{5}\sqrt{5}) < X_1 < 5. & (2.51b) \end{cases}$$

2.6.2. From system (2.51) we conclude that on the cone $\bar{D} := \{X \in \mathbb{R}^2_{++} \ / \ 3 \leqq X_1/X_2 \leqq 5\}$, the function F is a valid RP-function, see theorem 2.4, but does not exhibit the Law of Diminishing Returns to Factors.

NOTES

1. The contents of this chapter are based on JUNIUS (1976a), (1976b).
2. SHEPHARD (1970), chapter 2. Actually, SHEPHARD assumes the existence of n production factors, see also SHEPHARD (1953), chapter 1.
3. SHEPHARD (1967), p. 210.
4. SHEPHARD (1970), section 2.1.
5. Production for intermediate purposes is disregarded.
6. $\mathbb{R}^k_+ := \{X \in \mathbb{R} \ / \ X_i \geqq 0, \ i = 1,\ldots, k\}$ for each positive integer k.
7. RUYS (1975), p. 27.
8. SHEPHARD (1967), p. 218; RUYS (1975), p. 31.
9. SHEPHARD (1970), pp. 13, 14.
10. See for example SHONE (1975), chapter 5.
11. SHEPHARD (1970), pp. 14, 15.

12. SHEPHARD (1967), p. 219.
13. SHEPHARD (1974), p. 260.
14. PLASMANS (1975), chapter 1.
15. For a discussion of the CES-class, see chapter 4.
16. SHEPHARD (1970), p. 17.
17. SHEPHARD (1970), pp. 18, 19.
18. SHEPHARD (1970), p. 20.
19. SHEPHARD (1970), pp. 22, 23.
20. SHEPHARD (1970), p. 22.
21. SHEPHARD (1970), section 2.3 and section 2.4.
22. SHEPHARD (1970), pp. 24, 25.
23. FØRSUND (1974), p. 109.
24. SHEPHARD (1970), p. 29, DIEWERT (1976).
25. A function $u = g(z)$, $z \in \mathbb{R}^n$, is called homogeneous of degree h, if and only if for each $t \gtreqless 0$ and each z^o: $g(tz^o) = t^h g(z^o)$. For a discussion see UEBE (1976), pp. 54, 55.
26. SHEPHARD (1970), chapter 2, propositions 7, 8.
27. Super-additivity: $H(X + X') \gtreqless H(X) + H(X')$.
28. PLASMANS (1975), p. 22. In PLASMANS' definition a neoclassical production function is homogeneous and may generate unbounded efficient subsets.
29. $\mathbb{R}^k_{++} := \{X \in \mathbb{R}^k / X_i > 0, i = 1, .., k\}$.
30. For more traditional treatments see FERGUSON (1971), section 4.5., or HENDERSON, QUANDT (1958), p. 51.
31. See also SHEPHARD (1967), pp. 230-232.
32. This to distinguish this type of functions from more traditional views; for alternative views see KRELLE (1969), chapter 3.
33. PLASMANS (1975), p. 22.
34. SHONE (1975), p. 187.
35. FRISCH (1965), part two, FERGUSON (1971), part 1, KRELLE (1969), chapters 4, 5, MORISETT (1953).
36. VAZQUEZ (1971).
37. HICKS (1932), p. 289

38. FERGUSON (1971), part 1.
39. For alternative views see BORTS, MISHAN (1962), and KUIPERS (1970), page 27, figure 2.3.2. and its comments below.
40. SATO (1967), FØRSUND (1974), chapter 5.
41. JUNIUS (1976b).
42. For extensive discussions see EICHHORN (1970), chapter 2, SHEPHARD (1970), section 2.7, SHEPHARD-FäRE (1974).

    In the literature it is often postulated that a neoclassical production function F is homogeneous (or homothetic) with

    $F'_i > 0, F''_{ii} < 0, i = 1,2,$

    see for example VAZQUEZ (1971) and BECKMANN (1974). Def. 2.9 and the results of this subsection show that these inequalities must be regarded as too restrictive.

3. Neoclassical production functions in economic dynamics

3.1. INTRODUCTION

3.1.1. The concept of a neoclassical production function often arises as one of the basic ingredients in economic-*theoretical* models. Mostly, however, the production function is defined in a traditional way. This means that the postulated production function does not fit completely into the framework initiated by SHEPHARD. In this chapter we modify a number of theoretical studies from the field of economic dynamics by taking account of definition 2.9 of an S-neoclassical production function.

In section 3.2 the neoclassical one-sector growth model serves as a basis for the construction of a simple example how to investigate the consequences of the insertion of an RP-function into this model. In section 3.3 we modify a study of BECKMANN on neutral technical change by the insertion of an S-neoclassical production function into BECKMANN's framework.

3.1.2. The discussions in this chapter show that, on the acceptation of SHEPHARD's view of a valid production technology a number of additional comments can be made regarding the properties of the neoclassical growth model. In the same spirit, it is possible to extend and adapt studies on technical change and to reformulate some of the results which were obtained.

3.2 THE NEOCLASSICAL ONE-SECTOR GROWTH MODEL

3.2.1. As the first example how to investigate the differences between the consequences of SHEPHARD's view of production technologies and traditional neoclassical views, we take the well-known neoclassical one-sector growth model as a starting-point. In order to stress these differences, we omit in our presentation complications of technical change, depreciation etc. and just concentrate on the production conditions. E.g., BURMEISTER and DOBELL also show how to investigate the implications of several alternative specifications of production conditions (1). From SHEPHARD's point-of-view these specifications are nonvalid, since they are all of the CES-type.

3.2.2. The neoclassical one-sector growth model basically consists of four equations in which five variables appear (2). These five variables are assumed to be continuously differentiable functions of the variable t (= time). The four equations are:

$$Y_t = F(K_t, L_t), \tag{3.1a}$$

$$Y_t = C_t + I_t, \tag{3.1b}$$

$$\dot{K}_t = I_t, \quad K_o > 0, \tag{3.1c}$$

$$L_t = L_o \exp(gt), \quad L_o > 0, \quad g > 0, \tag{3.1d}$$

where Y := equilibrium output (= income), K := capital input (= capital stock available), L := labor input (= labor supply), I := gross investment (= net investment), C := consumption, $\dot{K} = dK/dt$ and g a constant.

3.2.3. Eq. (3.1a) states that output is produced by means of the two production factors capital and labor. In comparison with the usual assumptions, the production technology (function) in this growth model will now be

modified as follows:
1. It will be assumed that the production function F is a *restricted* production function (see def. 2.10); this implies that F is only defined on a non-empty closed convex cone $\bar{D} \subset \mathbb{R}^2_{++}$.
2. It will be assumed that on the cone $\bar{D}$ the function F is homogeneous of degree one with respect to K_t and L_t, i.e.:

$$F(K_t, L_t) = \hat{F}(K_t, L_t). \tag{3.2}$$

Since K and L are differentiable functions of t we may write:

$$dF = F'_K \dot{K} dt + F'_L \dot{L} dt, \quad \dot{L} := dL/dt, \tag{3.3}$$

i.e., if K and L do not change (dK = dL = 0), we conclude from eq. (3.3) that, if $dt \neq 0$:

$$F'_t = 0, \tag{3.4}$$

which implies that the postulated production technology (eq. (3.1a) does not show the property of technical change.

As noted by BURMEISTER, DOBELL (3), the implicit assumptions of full employment of capital and labor imply that each factor price equals its real marginal productivity. If each factor price is positive, then both marginal productivities are positive. So if SHEPHARD's view of a production technology is applied to the neoclassical growth model, only the restricted production function F can be considered, not the (unrestricted) S-neoclassical production function H.

Furthermore, let the capital-labor ratio k_t be defined by:

$$k_t := K_t/L_t \tag{3.5}$$

Then there exist scalars a and b such that:

$$0 < a < b < \infty, \quad (3.6a)$$

$$f(k_t) := F(k_t, 1), \text{ domain } f = [a,b] \subset \mathbb{R}_{++}, \quad (3.6b)$$

$$\text{on } (a,b): f > 0, \ 0 < f'k_t/f < 1, \ f'' < 0. \quad (3.6c)$$

Without loss of generality it will be assumed that:

$$\lim_{k_t \downarrow a} k_t f'/f = 1, \quad (3.7a)$$

$$\lim_{k_t \uparrow b} f' = 0. \quad (3.7b)$$

In figure 3.1 below we illustrate a graph of the function f in order to stress that:

1. In our neoclassical growth model *efficient* production is only possible on a proper subset in $\mathbb{R}^2_{++}$.
2. Domain f is the *largest possible* set for efficient production (see eqs. (3.7a) and (3.7b)).

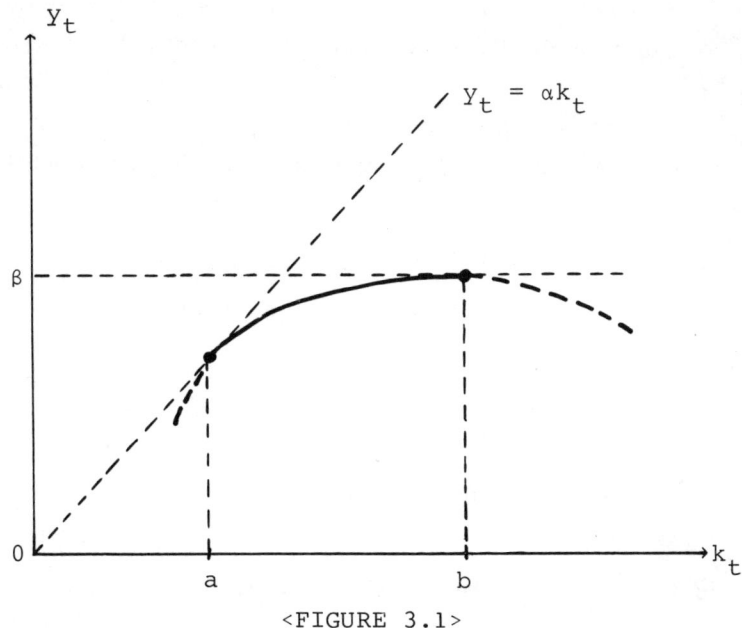

<FIGURE 3.1>

In figure 3.1 the half-rays $y_t = \alpha k_t$ and $y_t = \beta > 0$ restrict the graph of f on the interval [a,b]. These half-rays directly follow from eqs. (3.7a) and (3.7b). Eqs. (3.1b), (3.1c) and (3.1d) are well-known and their interpretations can be found in the references (4).

3.2.4. Let per capita consumption c_t be defined by:

$$c_t := C_t/L_t. \tag{3.8}$$

From $K_t = k_t L_t$, $\dot{L}_t/L_t = g$ and eq. (3.8) we obtain:

$$\dot{k}_t = f(k_t) - gk_t - c_t. \tag{3.9}$$

This differential equation is the *fundamental equation* of the neoclassical one-sector growth model. If c_t is a given function of either the variable k_t or the variable t, eq. (3.9) reduces to a differential equation in the unknown function k_t. The general solution of eq. (3.9) contains one integration constant, this constant can be related to a single initial condition of the form $k = \bar{k}_o$ for t = 0.

3.2.5. SOLOW extended the neoclassical growth model by the insertion of the following HARROD-DOMAR consumption function (5):

$$C_t = (1 - s)Y_t, \quad 0 < s < 1, \tag{3.10}$$

where the constant s is to be interpreted as the average propensity to save. In terms of per capita consumption we find that:

$$c_t = (1 - s)f(k_t). \tag{3.11}$$

The insertion of the right-hand side of eq. (3.11) into eq. (3.9) yields:

$$\dot{k}_t = sf(k_t) - gk_t = \varphi(k_t), \quad \dot{k}_t := dk/dt, \tag{3.12}$$

This equation states (6) that the growth rate of the capital-labor ratio is equal to 'the surplus available after equipping all new labor at the existing standard'. In figures 3.2a, 3.2b and 3.2c below we illustrated the three possibilities with respect to the graph of the function φ. The shaded area is used to stress the basic property of this neoclassical growth model that the growth rate in the capital-labor ratio, $\dot{k}_t$, may be strictly negative, fig. 3.2a, or strictly positive, fig. 3.2c. In figure 3.2b there exists a value $\underline{k}$ for which $\dot{k}_t = 0$, i.e. $\dot{K}_t = \dot{L}_t$, so-called balanced growth.

In order to study the properties of the fundamental equation (3.12) we now adopt three definitions (7):

DEFINITION 3.1: A value $\underline{k} \in [a,b]$ is called an *equilibrium value* of eq. (3.12), if $\varphi(\underline{k}) = 0$.

I.e., for each equilibrium value $\underline{k}$ we obtain that capital, labor and output are growing at the same rate, since $\dot{k}_t = \varphi(\underline{k}) = 0$ and $\dot{x}_t = f'(\underline{k})\dot{k}_t = 0$, $\dot{x}_t := dx/dt$.

DEFINITION 3.2: Eq. (3.12) is *globally stable*, if for any $k_o \in [a,b]$ there exists an equilibrium value $\underline{k}(k_o) \in [a,b]$, such that each solution $\pi_t(k_t)$ of eq. (3.12) has the property:

$$\lim_{t \to \infty} \pi_t(k_t) = \underline{k}(k_o). \tag{3.13}$$

Observe that in def. 3.2 it is crucial that the equilibrium value $\underline{k}$ may depend on the initial value k_o.

DEFINITION 3.3: An equilibrium value $\underline{k} \in [a,b]$ of eq. (3.12) is *locally asymptotically stable*, if for any $k_o \in [a,b]$ each solution $\pi_t(k_o)$ of eq. (3.12) has the property:

$$\lim_{t \to \infty} \pi_t(k_o) = \underline{k}. \tag{3.14}$$

If def. 3.3 it is crucial that reference is made to an equilibrium value $\underline{k}$ which does *not* depend on the initial value k_o.

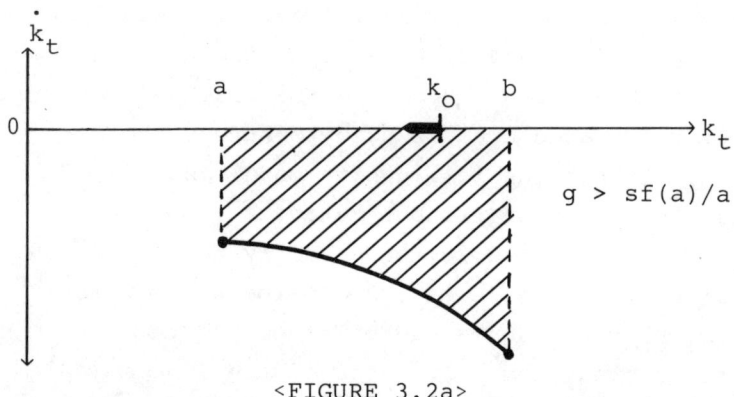

<FIGURE 3.2a>

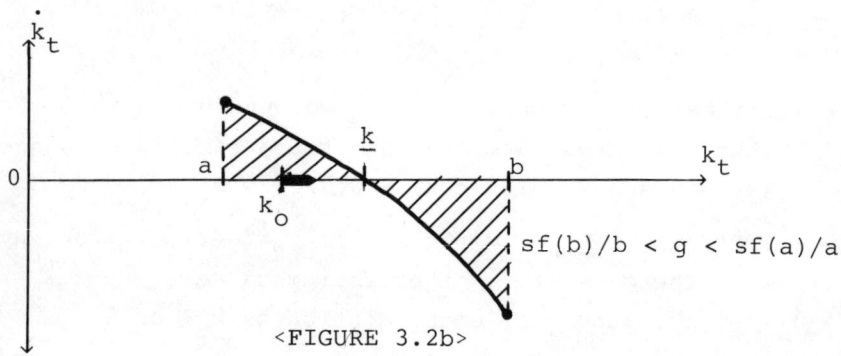

<FIGURE 3.2b>

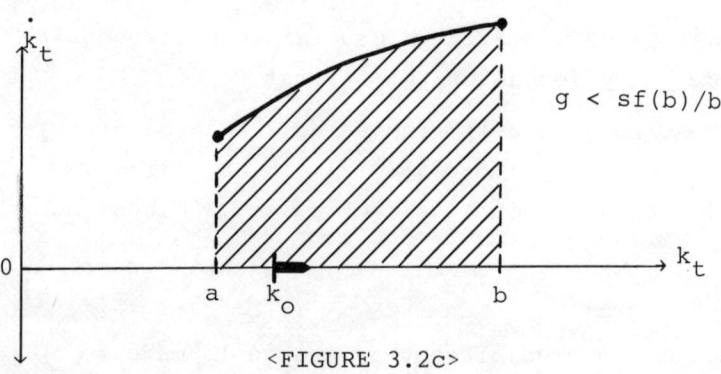

<FIGURE 3.2c>

We shall now investigate the properties of the (modified) neoclassical growth model (eqs. (3.1a),...,(3.1d), eq. (3.10)), by studying the graphs of the relevant function sketched in figure 3.2b.

From this figure we conclude that:

1. If $a < k_o < \underline{k}$, then $\dot{k}_t > 0$, i.e. the capital-labor ratio increases.
2. If $\underline{k} < k_o < b$, then $\dot{k}_t < 0$, the capital-labor ratio decreases.

Taking account of def. 3.3 yields the following proposition.

THEOREM 3.1: If:

$$sf(b)/b < g < sf(a)/a \qquad (3.15)$$

then there exists an equilibrium value $\underline{k}$ such that $\underline{k}$ is asymptotically stable in the region [a,b].

With the aid of def. 3.2 it can be easily proved that, if system (3.15) holds, the equilibrium value $\underline{k}$ is also globally stable in the region [a,b].

That is to say, in comparison with the results of the original model of SOLOW we find that the application of SHEPHARD's view, by means of the definition of an RP-function F, restricts the values for which balanced growth can occur regarding the relevant variables, i.e. the equilibrium output, capital input and labor input.

3.3. NEUTRAL TECHNICAL CHANGE AND NEOCLASSICAL PRODUCTION FUNCTIONS

3.3.1. As a second but more fundamental example of the implications of our approach we shall now consider in this section a recent study by BECKMANN concerning the relationship between different types of technical change and the implied properties of the production function (8). BECKMANN's analysis can be summarized as follows: assuming a certain type of neutral technical change BECKMANN *only* proved that this type must lead to a class of *homothetic* functions. It is our opinion, however, that BECKMANN's problem must be extended. Given a view of what must be considered as a valid production function, what conditions must be assumed regarding neutral technical change for neutral technical change to lead to a class of for example S-neoclassical production functions?

In subsection 3.3.3. the relationship between so-called generalized HICKS neutral technical change and the properties of the dynamic production function are analyzed. In subsection 3.3.4. we study generalized BECKMANN neutral technical change.

3.3.2. In order to be able to modify BECKMANN's study in terms of our arguments we shall introduce a number of definitions and assumptions. Some of them cannot be found explicitly in BECKMANN's paper, at least not in the form stated here. But we shall make it clear that BECKMANN used them implicitly, by referring to the corresponding places in his text.

It is assumed that a *dynamic* production function:

$$Y = H(K,L,t), \text{ domain } H = \mathbb{R}^3_+ \qquad (3.16)$$

is given, stating that the output rate Y depends on capital input K, labor input L and the variable t (= time). It is assumed that for $t^o \geqq 0$, t^o fixed, the function $H(K,L,t^o)$ is an S-neoclassical production function in the variables K and L (9). For $K^o > 0$ and $L^o > 0$, (K^o,L^o) fixed, the function $H(K^o,L^o,t)$ is continuously differentiable with respect to t and $H'_t > 0$. The last assumption states that the function H shows the property of *technical change*.

DEFINITION 3.4: Let H be a dynamic production function. The function H represents *output augmenting* technical change (10), if and only if there exists a function v from $\mathbb{R}_+^2$ to $\mathbb{R}_+$ and a function G from $\mathbb{R}_+^2$ to $\mathbb{R}_+$, such that for each $(K,L,t) \in$ domain H the function H can be written in the form:

$$H(K,L,t) = v(G(K,L),t). \qquad (3.17)$$

Observe that $v'_t > 0$ if H represents output augmenting technical change; i.e. for (K^o, L^o) fixed, the function v is strictly increasing in the variable t.

DEFINITION 3.5: Let H be a dynamic production function. The function H represents *output* and *capital augmenting* technical change (11), if and only if there exists a function v from $\mathbb{R}_+^2$ to $\mathbb{R}_+$, a function G from $\mathbb{R}_+^2$ to $\mathbb{R}_+$ and an increasing function q from $\mathbb{R}_+$ to $\mathbb{R}_+$ such that for each $(K,L,t) \in$ domain H the function H can be written in the form:

$$H(K,L,t) = v(G(q(t)K,L),t) \qquad (3.18)$$

Observe that $v'_t > 0$ if H represents output and capital augmenting technical change. From the differentiability of H with respect to the variable K we obtain that $q' > 0$. Output and labor augmenting technical change may be defined by interchanging the variables K and L in definition 3.5.

DEFINITION 3.6: Let H be a dynamic production function. A function s is a (dynamic) MRS- (= *marginal rate of substitution*) function (12), if and only if:

$$\begin{aligned} s(K,L,t) &= H'_K(K,L,t)/H'_L(K,L,t), \\ \text{domain } s(K,L,t) &= D^o, \end{aligned} \qquad (3.19)$$

where D^o is the set in $\mathbb{R}_{++}^3$ on which both marginal factor productivities are at least zero.

DEFINITION 3.7: Let H be a dynamic production function. A function σ is a (dynamic) ES- *(= elasticity of substitution)* function (13), if and only if:

$$\sigma(K,L,t) = -(d\log(K/L)/d\log s)_{Y=\text{const.}}, \quad (3.20)$$
$$\text{domain } \sigma(K,L,t) = D^o.$$

3.3.3. In this subsection we shall review and modify BECKMANN's analysis of the relationship between generalized HICKS neutral technical change (14) and the implied properties of the dynamic production function H. We define the concept of generalized HICKS neutral technical change as follows.

DEFINITION 3.8: Let H be a dynamic production function. The function H represents *generalized HICKS neutral technical change* (= g.H.n.t.c.), if and only if there exists a function $\hat{s}$ from $\mathbb{R}_{++}$ to $\mathbb{R}_{++}$, such that for each $(K,L,t) \in D^o$, the following relation holds:

$$s(K,L,t) = \hat{s}(k), \quad k := K/L. \quad (3.21)$$

Observe that in the definition of g.H.n.t.c. only technical considerations are playing a role. If g.H.n.t.c. holds, then:
1. The marginal rate of substitution is a function of the capital-labor ratio, $\hat{s}$ is homogeneous of degree zero with respect to K and L.
2. The marginal rate of substitution is independent of time.

If in addition the function H is homogeneous of degree one with respect to K and L, definition 3.8 reduces to HICKS neutral technical change.

BECKMANN obtained (16) a theorem concerning the relationship between g.H.n.t.c. and the properties of the production function H: if H represents g.H.n.t.c. then:
1. Technical change is output augmenting, see def. 3.4.

2. The function H is homothetic (in a neoclassical sense).

This theorem will now be modified and extended below.

THEOREM 3.2: Let $\hat{s}$ be a *given* function with domain $\hat{s} = (a,b) \subset \mathbb{R}_{++}$, on (a,b) the function $\hat{s}$ is continuously differentiable and $\hat{s} > 0$ and $\hat{s}' < 0$. If there exists a function $s(K,L,t)$ in such a way that for each $(K,L,t) \in D^o$, eq. (3.21) holds and if the function $F(K,L,t)$ is an arbitrary solution of differential equation (3.19), then:

1. There exists a function $A(t,Y)$ from $\mathbb{R}_+^2$ to $\mathbb{R}_+$, A continuously differentiable on $\mathbb{R}_+$ with $A'_t < 0$ and $A'_Y > 0$.
2. There exists a function $\hat{F}(K,L)$ from a closed convex cone $\overline{D}$ to $\mathbb{R}_+$, $\hat{F}$ homogeneous of degree one with respect to K and L.
3. The functions A and $\hat{F}$ are defining the solution F as an implicit function in such a way that on the cone $\overline{D}$:
   - for t fixed the function F is an RP-function of an S-neoclassical production function,
   - the function F represents g.H.n.t.c. and
   - the function F represents output augmenting technical change.

This reconsideration of the theorem obtained by BECKMANN has two main characteristics. First, if the function $\hat{s}$ is an arbitrary mathematical function with the properties of a static MRS-function, the additional restrictions are presented such that actually the properties of the dynamic production function fulfil certain criteria of a production technology (in our frame-work: F is an RP-function). Secondly, the results of BECKMANN's study are generalized in such a way that his approach fits into the concept of S-neoclassical production functions.

This last statement follows from the fact that for t fixed each RP-function F can be extended to an S-neoclassical production function by the application of eqs. (2.11c),...,(2.11e).

Proof: Theorem 5.2 will be proved by reviewing BECKMANN's analysis (17) and will consist of five steps. In step 1 we shall rewrite differential eq. (3.20). In step 2 we shall obtain a mathematical expression for an arbitrary solution of this equation. In step 3 it will be shown that this solution defines the output rate as an implicit function F of the variables K, L and t. In step 4 and 5 it will be shown that the function F represents output augmenting technical change and that for t fixed the function F is an RP-function of an S-neoclassical production function.

1) Let t and Y be fixed and let $F(K,L,t)$ be a solution of differential equation (3.20). Along each isoquant $I_F(Y)$ of F (see def. 2.5) we obtain ($dY = dt = 0$):

$$dF = F'_K dK + F'_L dL = 0. \tag{3.22}$$

From eqs. (3.20) and (3.21) we find after a number of elementary operations that for $K/L \in (a,b)$:

$$dK/dL = -1/\hat{s}(K/L). \tag{3.23}$$

The insertion of $K = kL$ and $dK/dL = k + Ldk/dL$ into eq. (3.23) yields:

$$dL/L = -\hat{s}(k)dk/(k\hat{s}(k) + 1). \tag{3.24}$$

2) Eq. (3.24) is a differential equation in which the variables L and k have been separated. Since it is assumed that Y is a function of K, L and t, it follows that each solution of eq. (3.24) contains an integration-constant A (18) which must be taken as a strictly positive function of the variables t and Y. We write successively:

$$\ln L = -\int \hat{s}(k)dk/(k\hat{s}(k) + 1) + \ln A(t,Y), \qquad (3.25)$$

$$A(t,Y) = L\exp\int \hat{s}(k)dk/(k\hat{s}(k) + 1), \qquad (3.26)$$

$$A(t,Y) = L\exp\int \hat{u}(k), \text{ domain } \hat{u} = [a,b], \qquad (3.27)$$

$$A(t,Y) := L\hat{f}(k), \text{ domain } \hat{f} = [a,b], \qquad (3.28)$$

$$A(t,Y) := \hat{F}(K,L),$$
$$\text{domain } \hat{F} = \bar{D} := \{(K,L) \in \mathbb{R}^2_{++}/\ K/L \in [a,b]\}, \qquad (3.29)$$

where the function $\hat{F}$ is independent of t and is homogeneous of degree one with respect to K and L. Observe that domain of $\hat{F}$ is *independent* of the value of the variable t (see eq. (3.29)).

3) Consider the relation $T \subset \mathbb{R}^3$ defined by:

$$T(t,Y,Z) := A(t,Y) - Z = 0 \qquad (3.30)$$

where $Z := \hat{F}(K,L)$; the variable Z may be interpreted as a time-independent index of K and L by using the function $\hat{F}$. Relation (3.30) defines the variable Y as an implicit function of the variables Z and t, if (19):

$$T'_Y = A'_Y \neq 0. \qquad (3.31)$$

But (3.31) holds by assumption, since if $A'_Y = 0$, this would imply that A is not a function of Y. I.e. there exists a function v from $\mathbb{R}^2_{++}$ to $\mathbb{R}_{++}$:

$$Y = v(Z,t) = v(\hat{F}(K,L),t) = F(K,L,t). \qquad (3.32)$$

4) Eq. (3.29) yields for dK = dL = 0:

$$A'_t dt + A'_Y dY = 0, \qquad (3.33)$$

or

$$dY/dt = -A'_t/A'_Y = v'_t \qquad (3.34)$$

(see eq. (3.32)).
For dt = 0, we conclude from eq. (3.29) and $Z = \hat{F}(K,L)$:

$$A'_Y dY = dZ, \qquad (3.35)$$

or

$$dY/dZ = 1/A_Y'. \qquad (3.36)$$

Using eqs. (3.34) and (3.36), we conclude that: if A is continuously differentiable on $\mathbb{R}^2_{++}$ and $A_t' < 0$ and $A_Y' > 0$, then on the cone $\bar{D}$ (see eq. (3.29)):

1. For t fixed the function F is homothetic in a neoclassical sense.
2. The function F represents output augmenting technical change.
3. By assumption the function F represents g.H.n.t.c.

5) In the last step we shall investigate the properties of the function $\hat{f}$ which enters into eq. (3.28). This function will determine whether the function $F(K,L) = v(\hat{F}(K,L),t^o)$, for t^o fixed, see eq. (3.32), represents an RP-function of a static S-neoclassical production function. Since the function $\hat{s}$ has the properties of a static MRS-function it follows directly from the discussions in subsection 3.5.3. (see theorem 2.5) that $\hat{f}$ is an RCI-function. This implies that F is actually an RP-function of a static S-neoclassical production function. Q.E.D.

That is to say, we have shown that it is possible to introduce additional restrictions on the concept of g.H.n.t.c. so that this type of technical change leads to output augmenting technical change and to an RP-function in the sense of our framework of chapter 2. The RP-function F (see eq. (3.32) can be extended to a (static) S-neoclassical production function. This implies that to each solution of differential equation (3.22) we can attribute a dynamic production function.

It can be easily proved that theorem 3.2 can be reversed in the following way: if the function H is a dynamic production function and represents output augmenting technical change, then the function H represents g.H.n.t.c. and on an interval $(a,b) \subset \mathbb{R}_{++}$ the function $\hat{s}$ has the properties of a static MRS-function.

3.3.4. In this subsection we shall review and modify BECKMANN's analysis on the relationship between generalized BECKMANN neutral technical change and the implied properties of the postulated dynamic production function (20). In order to define the concept of generalized BECKMANN neutral technical change we shall first give the definition of the relative factor shares (21).

DEFINITION 3.9: Let r and w denote the market prices for capital and labor. Let the output market-price be unity, H be a dynamic production function and let it be assumed that:

$$r = H'_K, \quad w = H'_L, \qquad (3.37)$$

for each $(K,L,t) \in D^o$, i.e., on D^o the marginal productivity conditions hold.
Then the function β from D^o to $\mathbb{R}_{++}$ is a function of the *relative factor shares*, if and only if:

$$\beta(K,L,t) = KH'_K(K,L,t)/LH'_L(K,L,t). \qquad (3.38)$$

The function β is one of the ingredients for the definition of generalized BECKMANN neutral technical change (22).

DEFINITION 3.10: Let H be a dynamic production function. The function H represents generalized BECKMANN neutral technical change (:= g.B.n.t.c.), if and only if there exists a function $\hat{\sigma}$ from $\mathbb{R}_{++}$ to $\mathbb{R}_{++}$ such that for each $(K,L,t) \in D^o$ the following relation holds:

$$\sigma(K,L,t) = \hat{\sigma}(\beta(K,L,t)). \qquad (3.39)$$

From this definition it follows that g.B.n.t.c. takes

the ES-function of a dynamic production function H as a function of the relative factor shares only. If the function H is homogeneous of degree one, KRELLE calls this type of technical change BECKMANN neutral technical change (23). Observe that by means of def. 3.10 technical and economic considerations enter in the concept of g.B.n.t.c., while the concept of g.H.n.t.c. is only based on technical considerations (24).

Observe that the manner of defining relative factor-shares, (see def. 3.9) is crucial. This definition, in combination with the definition of generalized BECKMANN neutral technical progress, yields in BECKMANN's theorem *capital* and *not* labor augmenting technical progress.

BECKMANN derived the following theorem (25). If the function H represents g.B.n.t.c., then:
1. Technical change is output and capital augmenting, see def. 3.5.
2. The function H is homothetic in a neoclassical sense.

This theorem will now be modified and extended as follows:

THEOREM 3.3: Let $\hat{\sigma}$ be a *given* function with domain $\hat{\sigma} = (a,b) \subset \mathbb{R}_{++}$, $\hat{\sigma}$ continuous on (a,b) and on (a,b):

$$\hat{\sigma} > 1. \qquad (3.40)$$

If there exist a function $\sigma(K,L,t)$ such that for each $(K,L,t) \in D^o$, eq. (3.39) holds and if the function F is a solution of differential equation (3.38), then:
1. There exists a function $A(t)$ from $\mathbb{R}_+$ to $\mathbb{R}_+$, A continuously differentiable on $\mathbb{R}_{++}$ with $A' > 0$.
2. There exists a function $B(t,Y)$ from $\mathbb{R}_+^2$ to $\mathbb{R}_+$, B continuously differentiable on $\mathbb{R}_{++}^2$ with $B_t' > 0$ and $B_Y' > 0$.
3. There exists a function $\hat{f}$ with domain a closed inter-

val $[a,b] \subset \mathbb{R}_{++}$ and $\hat{f}$ an RCI-function on $[a,b]$.
4. The functions A, B and $\hat{f}$ define the solution F of eq. (3.38) as an implicit function in such a way that on a closed convex cone $\overline{D}$:
   a. For t fixed the function F is an RP-function.
   b. The function F represents output and capital augmenting technical change.
   c. The function F represents g.B.n.t.c.

We first discuss the characteristics of this reconsideration of the second theorem of BECKMANN. Additional restrictions are again introduced so that the implied properties of the solution F fits into a view of production technologies, i.e. the solution F as an RP-function. To guarentee the existence of isoquants which are convex to the origin, it must be assumed that the function $\hat{\sigma}$ exceeds the value one. If it would be assumed that $0 < \hat{\sigma} \leq 1$, then it follows that for t fixed the MRS-function $\hat{s}$ exhibits the properties $\hat{s} > 0$, $\hat{s}' > 0$.

Proof: Like in the proof of theorem 5.2., BECKMANN's way of analyzing (26) will be taken as a starting-point, our proof will be decompounded into six steps. In step 1 we shall obtain a differential equation on the basis of eq. (3.39). In step 2 it is shown that each solution of this differential equation yields the capital-labor ratio as a function of the relative factor shares. In step 3 the properties of this solution are investigated. In step 4 a second differential equation will be obtained by considering eq. (3.38). In step 5 it will be shown that each solution of this differential equation yields F as an implicit function. In step 6 we shall investigate the properties of this function.

1. Using $k := K/L$, defs. 3.6 and 3.9, yields:

$$\beta = ks(K,L,t) \tag{3.41}$$

and if dt = 0 we find:

$$d(\log\beta)/d(\log k) = 1 + d(\log s)/d(\log k). \quad (3.42)$$

From ineq. (3.40) it follows that:

$$d(\log\beta)/d(\log k) = 1 - 1/\hat{\sigma}(\beta), \quad (3.43)$$

which can be rewritten as:

$$dk/k = d\beta/\beta(1 - 1/\hat{\sigma}(\beta)). \quad (3.44)$$

2. A function k is a solution of differential eq. (3.44) if there exists a strictly positive function A(t) in such a way that we may write (in terms of indefinite integrals):

$$\log A(t)k = \int d\beta/\beta(1 - 1/\hat{\sigma}(\beta)), \quad (3.45)$$

or:

$$k_t = \varphi(\beta), \text{ domain } \varphi = (a,b), \quad (3.46)$$

where:

$$k_t := A(t)k, \quad \varphi(\beta) := \exp\int d\beta/\beta(1 - 1/\hat{\sigma}(\beta)). \quad (3.47)$$

Eq. (3.46) states that each solution of differential eq. (3.44) defines the capital-labor ratio as a function of the relative factor shares.

3. In this step we shall investigate the properties of the function φ. From $\beta > 0$, $(1 - 1/\hat{\sigma}(\beta)) > 0$ and:

$$\exp\int d\beta/\beta(1 - 1/\hat{\sigma}(\beta)) > 0 \quad (3.48)$$

it follows that the *range* of the function φ is an open interval in $\mathbb{R}_{++}$, say $k_t \in (\lambda_t, \mu_t)$. From this property and $k = k_t/A_t$ it follows that the boundaries of the variable k are *time-dependent*.

From:

$$d\varphi/d\beta = \varphi/\beta(1 - 1/\hat{\sigma}(\beta)) > 0 \quad (3.49)$$

we conclude that to the function φ we can attribute its inverse function $\Psi = \varphi^{-1}$ in such a way that:

$$\beta = \Psi(k_t) > 0, \text{ domain } \Psi = (\lambda_t, \mu_t), \qquad (3.50)$$

while from (3.49) we find that:

$$0 < k_t d\Psi/\Psi dk_t < 1. \qquad (3.51)$$

The inverse function Ψ and its implied properties will in the next three steps be the basis for the analysis concerning the question whether the solution F of differential equation (3.38) is an RP-function for t fixed. But first, by way of an interlude, two additional comments on the function Ψ will be made. First, let the function $\Psi^o(k) := \Psi(A(t^o)k)$, t^o fixed. Then the domain of the function Ψ^o is *time-dependent*. This means that the choice of the value t^o may influence the values of the boundaries of domain Ψ^o.

Secondly, the properties of the function Ψ^o can be used to clarify the implications of the assumption that on (a,b) the function $\hat{\sigma} > 1$. For t^o fixed we obtain:

$$s = \beta/k = \Psi^o(k)/k > 0. \qquad (3.52)$$

The function s in (3.52) is a static MRS-function if:

$$s'(k) = ((\Psi^o)'k - \Psi^o)/k^2 < 0. \qquad (3.53)$$

The inequality in the most right-hand side of (3.53) holds, if and only if ineqs. (3.51) holds. But ineqs. (3.51) holds, if and only if $\hat{\sigma}(\beta) > 1$. This means that the function s in eq. (3.52) has not the properties of an MRS-function if $0 < \hat{\sigma}(\beta) \leqq 1$ on (a,b), since $s' > 0$ and $s' > 0$ lead to the result that the isoquants of F are concave to the origin.

4. From eqs. (3.38) and (3.50) we obtain after rearranging the following differential equation:

$$KH'_K - \Psi(A(t)K/L)LH'_L = 0. \qquad (3.54)$$

BECKMANN obtains (27) the general solution of eq. (3.54) by considering the characteristic equation:

$$dK/dL = -K/L \Psi(A(t)K/L). \tag{3.55}$$

From $K = k_t L/A(t)$ and $dK/dL = (k_t + dk_t/dL)/A(t)$, it follows after substituting and rearranging, that eq. (3.55) can be written as:

$$dL/L = -\Psi(k_t) dk_t / (k_t \Psi(k_t) + k_t). \tag{3.56}$$

For $dY = dt = 0$ we get from eq. (3.56) that the general solution may be written as:

$$\log L + \log \hat{f}(k_t) = \log B(Y,t), \tag{3.57}$$

or:

$$B(t,Y) = L\hat{f}(k_t), \tag{3.58}$$

where:

$$\hat{f}(k_t) := \exp \int \Psi(k_t) dk_t / (k_t + k_t \Psi(k_t)), \tag{3.59}$$
$$\text{domain } \hat{f} = (\lambda_t, \mu_t),$$

and $B(t,Y)$ is a strictly positive function of t and Y on $\mathbb{R}^2_+$.

5. Consider the relation $T \subset \mathbb{R}^3$ defined by:

$$T(t,Y,Z) := B(t,Y) - Z = 0, \tag{3.60}$$

where $Z := L\hat{f}(k_t)$. This relation defines the variable Y as an implicit function of the variables Z and t, if:

$$T'_Y = B'_Y \neq 0, \tag{3.61}$$

which holds by assumption. I.e., we may write:

$$Y = v(L\hat{f}(A(t)K/L),t) = v(\hat{F}(A(t)K,L),t) = \tag{3.62}$$
$$= F(K,L,t).$$

If $B'_t > 0$, $B'_Y > 0$ and $A' > 0$, the function v *represents output and capital augmenting technical progress and is homothetic with respect to K and L for t fixed*. More par-

ticularly, $\hat{F}$ is homogeneous of degree one with respect to K and L. By assumption F represents g.B.n.t.c.
6. Let for t^o fixed the function F^o be defined by

$$F^o(K,L) := v(\hat{F}(A(t^o)K,L),t^o), \qquad (3.63)$$

with domain F^o a non-empty open convex cone in $\mathbb{R}^2_{++}$ which depends on the chosen value t^o, i.e. is time-dependent. We shall now investigate whether F^o represents an RP-function of a static S-neoclassical production function.
From:

$$\hat{f}(k_t) > 0, \qquad (3.64a)$$

$$0 < k_t\hat{f}'(k_t)/\hat{f}(k_t) = k_t\psi(k_t)/(k_t\psi(k_t)+k_t)) < 1, \quad (3.64b)$$

$$\hat{f}''(k_t)/\hat{f}(k_t) = (k_t\psi'(k_t)-\psi(k_t))/(k_t\psi(k_t)+k_t)^2, (3.64c)$$

we find that (for t^o fixed) the function $\hat{f}$ may be interpreted as an RCI-function since it follows from (3.51) and (3.64c) that:

$$\hat{f}''(k_t) < 0. \qquad (3.65)$$

That is to say, F^o represents an RP-function of a S-neoclassical production function. Q.E.D.

Def. 2.9 can be used to extend the RP-function F^o in such a way that its extension, say H^o, represents a (static) S-neoclassical production function. This implies that the modified concept of g.B.n.t.c. generates dynamic production functions which fit into SHEPHARD's view. In the preceding analysis we reviewed the concept of g.B.n.t.c. This mixture of technical and economic considerations, together with the basic assumption that the elasticity of substitution must exceed the value one, leads to the following interpretation. If the marginal productivity conditions hold, then we may write (28):

$$\sigma = d(\log k)/d(\log \omega) > 1, \qquad (3.66)$$

where $\omega = H'_L/H'_K$, the relative factor price; from this inequality we see that the concept of generalized BECKMANN neutral technical progress can only be applied if the economy has the property that, if the relative factor price ω falls (increases) by one percent, this fall (increase) is exceeded by the induced fall (increase) in the capital-labor ratio, since from eq. (3.66) we conclude:

$$dk/k > d\omega/\omega. \qquad (3.67)$$

This inequality elucidates the conclusion that g.B.n.t.c. can only be applied to a subclass of aggregate economies.

NOTES

1. BURMEISTER, DOBELL (1970), chapt. 2, sect. 2-4.
2. See BURMEISTER, DOBELL (1970), chapt. 2, TAKAYAMA (1974), chapt. 5.
3. BURMEISTER, DOBELL (1970), sect. 2-2, TAKAYAMA (1974), p. 433.
4. BURMEISTER, DOBELL (1970), without depreciation; TAKAYAMA (1970) with depreciation.
5. SOLOW (1956).
6. BURMEISTER, DOBELL (1970), p. 24, 25.
7. BURMEISTER, DOBELL, p. 27, 28.
8. BECKMANN (1974). This paper generalizes several results obtained in SATO, BECKMANN (1968).
9. BECKMANN only assumes $H'_i > 0$, $H''_{ii} < 0$, $i = K,L$, BECKMANN (1974), p. 6; see also STEHLING (1974), p. 67, who assumes that there is an input-combination (K^o, L^o) such that $H(K^o, L^o, t)$ strictly monotone for $t \gtreqless 0$.
10. BECKMANN (1974), p. 14. For alternative definitions see STEHLING (1974), p. 69 and EICHHORN, KOLM (1974), p. 37.

11. BECKMANN (1974), p. 18.
12. BECKMANN (1974), p. 12.
13. BECKMANN (1974), p. 12.
14. BECKMANN (1974), p. 13.
15. STEHLING (1974), p. 68.
16. BECKMANN (1974), p. 15.
17. BECKMANN (1974), pp. 13-15.
18. For the sake of clarity an integration constant A is added to the general solution, although the presentation will be given in terms of indefinite integrals.
19. TAKAYAMA (1974), p. 165, note 8.
20. BECKMANN (1974), p. 15.
21. BECKMANN (1974), p. 10.
22. BECKMANN (1974), p. 15; for an alternative definition of BECKMANN neutral technical change see EICHHORN, KOLM (1974), p. 43.
23. KRELLE (1969), p. 136.
24. BECKMANN (1974), p. 5, argues: 'it is (...) indispensable for this analysis to assume some version of the marginal productivity theory, if the relationships are to have any empirical significance'.
25. BECKMANN (1974), p. 18.
26. BECKMANN (1974), pp. 15-18.
27. BECKMANN (1974), p. 17; for applications of characteristic equations in the theory of production, see FØRSUND (1974), chapter V.
28. BURMEISTER, DOBELL (1970), section 1-4.

4. The indirect specification of classes of neoclassical production functions

4.1. INTRODUCTION

4.1.1. In chapter 3 we showed that it is possible to apply the concept of an S-neoclassical production function to several economic-theoretical studies. The purpose of the investigations in chapters 4 and 5, is to show that this concept can also be the basis for studies with a more *empirical* character. This is achieved by investigating:
1. The *way* by which a number of functional forms have been introduced in the literature to represent (traditional) two-factor production functions.
2. The properties of these functional forms.

With regard to the way by which functional forms are introduced we distinguish in this study *indirect* specification and *direct* specification of functional forms. Indirectly specified functional forms will be discussed in this chapter, directly specified functional forms in chapter 5. It must be noted that the specification of functional forms are an *essential* step in empirical analyses, because it precedes empirical evaluating.

With regard to the properties of each functional form is will be asked whether *(i) this functional form generates S-neoclassical production functions, (ii) such that the boundaries of the largest possible efficient subsets depend on the values of the parameters.* If for a specific functional form the answer to this question is in the affirmative, then this result suggests:

1. The possibility to use this functional form for estimating its parameters.
2. Subsequently, to test whether or not to reject the assumption of: largest possible efficient subsets which are bounded in $\mathbb{R}^2_{++}$.

4.1.2. In section 4.2. we start on discussing two methods for specifying indirectly the functional form of the class of CES production functions. These two methods are in section 4.3. the basis for an outline of specifying indirectly VES production functions. In section 4.4. we discuss a class of VES production functions which has been proposed by BRUNO and by VAZQUEZ.

The assumptions underlying the different indirectly specified functional forms are summarized in subsection 4.4.4. of this chapter. The table in this subsection elucidates how the assumptions in the two CES-class-generating methods can be altered in order to obtain the functional form for classes of VES production functions.

4.2. THE INDIRECT SPECIFICATION OF THE CLASS OF CES PRODUCTION FUNCTIONS

4.2.1. In many empirical studies concerning the search for functional forms for production functions, the class of CES production functions played and is still playing a prominent role (1). This class arises also in the field of consumer analysis and of aggregation theory (2). The CES-class cannot be a subclass of the framework of SHEPHARD, because this class generates largest possible efficient subsets which are *not* bounded in $\mathbb{R}^2_{++}$. The CES-class, however, is interesting from another viewpoint: it has been specified *indirectly* and its method of specification has given rise to the specification of a number of different class of production functions. By *indirect specification* the following is meant: the point of departure is, instead of assuming some functional form itself specified, it is assumed that this functional form *exhibits a property with regard to at least one of the partial derivatives*. Integration-theory then can be used to obtain the explicit functional form. Stated differently, the functional form is the outcome of an explicit theoretical model.

4.2.2. The indirect specification of the CES-class goes back to two groups of investigators, (i) BROWN and DE CANI and (ii) ARROW, CHENERY, MINHAS and SOLOW. Along different lines (3) both groups obtained independently this famous class. The results of their investigations show clearly the importance of the applications of the indirect specification method, the generalization and unification of a number of classes of production functions (4). E.g., one the basic constraints of the well-known COBB-DOUGLAS class, unitary elasticity of substitution, has been released.

4.2.3. In chapter 2 of this study, subsection 2.5.4., we found that if $\hat{\sigma}$ is a given function, then we may write for the implied class of MRS-functions:

$$\hat{s}(x) = \exp -\int dx/x\hat{\sigma}(x). \tag{4.1}$$

If $\hat{s}$ is an arbitrary function from this class, we obtain, see subsection 2.5.3.:

$$\hat{f}(x) = \exp \int dx/(x + 1/\hat{s}(x)), \tag{4.2}$$

or from eqs. (4.1) and (4.2):

$$\hat{f}(x) = \exp \int \frac{dx}{x + \exp\int dx/x\hat{\sigma}(x)}. \tag{4.3}$$

Eq. (4.3) has been derived by SATO (6) and can be used to obtain functional forms for classes of production functions, if $\hat{\sigma}$ is some given function. We shall use eq. (4.3) for discussing the B.DC- (= BROWN, DE CANI) method of obtaining the functional form of the class of CES RCI-functions (7). They assumed that:

$$\hat{\sigma}(x) = \sigma > 0, \text{ domain } \hat{\sigma}(x) = \mathbb{R}_{++}, \qquad (4.4)$$

i.e. a constant elasticity of substitution function. From eq. (2.33) it follows that this assumption is a basis for *indirectly specifying* a functional form, because this assumption states that a relation exists between a number of partial derivatives. The insertion of (4.4) into eq. (4.3) yields:

$$\hat{f}(x) = \exp \int dx/(x + Bx^{1/\sigma}), \qquad (4.5)$$

with B a strictly positive integration constant. The last integral can be solved by using the substitution:

$$x = t^{\sigma/(1-\sigma)}, \ \sigma \neq 1, \qquad (4.6)$$

from which we easily obtain:

$$\int dx/x + Bx^{1/\sigma}) = \sigma/(\sigma - 1) \log(t/(1 + B_1 t)), \qquad (4.7)$$

or

$$\hat{f}(x) = A_1 (x^{(\sigma-1)/\sigma} + B_1)^{\sigma/(\sigma-1)}, \qquad (4.8)$$

with A_1, B and B_1 strictly positive integration constants. The substitution of the value $\sigma = 1$ into eq. (4.3) yields:

$$\hat{f}(x) = \exp \int dx/(B + 1)x = Ax^{1/(B+1)}. \qquad (4.9)$$

That is to say, necessary and sufficient for a homothetic two-factor production function to have a constant ES-function is that the functional form of the RCI-function is given either by eq. (4.8) or by eq. (4.9).

Note that the B.DC.-method is not based on any additional 'economic assumption', e.g. market behavior (8).

Their method is based on purely technical information by means of (i) the homotheticity assumption and (ii) the assumption of constant elasticity of substitution. This means that in the B.DC.-method the production technology is regarded as a *datum*.

If we take it for granted that in the B.DC.-method the domain of the function $\hat{\sigma}$ must be taken as $\mathbb{R}_{++}$, then this yields that the largest possible cone for the RP-function:

$$\hat{F}(X) = A_1(X_1^{(\sigma-1)/\sigma} + B_1 X_2^{(\sigma-1)/\sigma})^{\sigma/(\sigma-1)}, \quad (4.10)$$

or for the COBB-DOUGLAS case:

$$\hat{F}(X) = A X_1^{1/(B+1)} X_2^{B/(B+1)}, \quad (4.11)$$

is $\mathbb{R}_{++}^2$. This implies that the results of the B.DC.-method do not fit into our approach. With regard to the value of the constant B_1 it can be proved that (9):

1. If the constant $\sigma \gtreqless 1$, then both production factors are non-essential.
2. If the constant $\sigma < 1$, then these are essential.
3. If the constant σ approaches to 0, then we obtain the class of WALRAS-LEONTIEF production functions.
4. If the constant σ approaches to ∞, then we obtain the class of linear production functions.

But regardless the value of the constant σ, it is clear that the B.DC.-method only yields RCI-functions which fit in the definition of an S-neoclassical production function, if it is assumed ad hoc that there are constants a and b such that the function $\hat{\sigma}$ is constant on the bounded interval $(a,b) \subset \mathbb{R}_{++}$.

4.2.3. The assumptions underlying the A.C.M.S.- (= ARROW, CHENERY, MINHAS, SOLOW) method for obtaining the functional form of the class of *homogeneous* CES production functions (10) are:

1. The existence of a homogeneous production function

$Y = F(X_1, X_2)$.
2. Output is numéraire.
3. The markets for the output and the first production factor are competitive, i.e.:

$$r_1 = F_1', \qquad (4.12)$$

with r_1 = the market-price for the use of one unit of the first production factor.

4. The *empirical* relation (ACMS-group: c = 1):

$$\ln Y = \ln a + b \ln r_1 + c \ln X_1. \qquad (4.13)$$

Assumptions 2, 3, 4 may be considered as economic assumptions. Eq. (4.12), an assumption concerning a partial derivative, is one of the ingredients for *specifying indirectly* the CES-class. The insertion of (4.12) into (4.13) yields a differential equation in which the function F is unknown. The general solution of this differential equation can be written as:

$$F(X) = (A_1(X_1^{(\sigma-1)/\sigma} + X_2^{(\sigma-1)/\sigma})^{\nu\sigma/(\sigma-1)}, \qquad (4.14)$$

or:

$$F(X) = (AX_1^{1/(B+1)} X_2^{B/(B+1)})^\nu \qquad (4.15)$$

with $\nu > 0$ the degree of homogeneity. This means that:

1. The A.C.M.S.-method also yields a class of production functions with the same functional form for the implied class of RCI-functions, eqs. (4.8), (4.9).
2. Again the function of the elasticity of substitution is constant with largest possible domain $\mathbb{R}_{++}$, which contradicts the approach of chapter 2.

4.2.4. In the preceding subsections we discussed two methods for obtaining the functional form of the class of CES production functions. From this discussion we conclude that the B.DC.-method is more general than the A.C.M.S.-method. First we see that the B.DC.-method results into a class of *homothetic* production functions (see also theorem 2.6), while the A.C.M.S.-method yields a class of *homogeneous* production functions. Secondly, in the A.C.M.S.-method a number of additional economic assumptions makes that the production technology is not regarded completely as a datum. This last statement can be elucidated in a different manner. Namely, it can be shown that the empirical relation (4.13) is implied by the following assumptions:

1. Let a production function $Y = F(X_1, X_2)$ exist.
2. Let this function be homogeneous of degree $\lambda > 0$.
3. Let the ES-function $\hat{\sigma}$ of the function F be constant on $\mathbb{R}_{++}$.
4. Let output be numéraire.
5. Let the output market and the market for the first production factor be competitive.

The first three assumptions lead to the following differential equation (11):

$$\lambda \sigma F''_{12}/F'_1 = (1 - (1-\lambda)\sigma) F'_2/F, \qquad (4.16)$$

or:

$$\lambda \sigma \frac{\ln F'_1}{\partial X_2} = (1 - (1-\lambda)\sigma) \frac{\partial \ln F}{\partial X_2}. \qquad (4.17)$$

The general solution of eq. (4.17) can be written as:

$$\ln F = b \ln F'_1 + c_1 \ln C(X_1), \qquad (4.18)$$

with $b := \lambda\sigma/(1 - (1-\lambda)\sigma)$, $c_1 := 1/(1 - (1-\lambda)\sigma)$, $\sigma \neq 1/(1-\lambda)$ and C a strictly positive function of the variable X_1. Assumptions 2, 4 and 5 then yield the empirical relation (4.13).

4.3. THE INDIRECT SPECIFICATION OF CLASSES OF VES PRODUCTION FUNCTIONS

4.3.1. In subsection 4.3.2. we give an outline of the properties of functional forms for classes of production functions which have been obtained by altering the basic assumption in the B.DC.-method. In subsection 4.3.3. functional forms will be reviewed which are based on alterations in the assumptions of the A.C.M.S.-method. This section concludes with a discussion of a class of production functions introduced by SATO and HOFFMANN and which has been obtained by means of a method that may be regarded as laying between the two CES-class-generating methods, which have been discussed before.

None of the functional forms to be reviewed in this section have the properties (i) that each functional form has largest possible efficient subsets which are bounded in $\mathbb{R}^2_{++}$ and (ii) that both boundaries of these subsets depend on the values of the parameters of this functional form. Like the CES-class, these functional forms represent classes of S-neoclassical production functions, if for these forms at least one boundary of each efficient subset is given ad hoc. For example this means that, the indirectly specified functional form itself cannot be used to test the assumption of bounded largest possible efficient subsets.

4.3.2. By assuming that an unknown production function F is homogeneous of degree one and that F has the following ES-function:

$$\hat{\sigma}(x) = 1 + bx, \qquad (4.19)$$

with:

$$\text{domain } \hat{\sigma} = \begin{cases} \mathbb{R}_{++}, & \text{if } b \geq 0, \\ (0,-1/b), & \text{if } b < 0, \end{cases} \quad (4.20)$$

SATO and REVANKAR independently obtained the following functional form for the implied class of production functions (12):

$$F(X) = AX_1^{1/(1+B)} (X_2 + (b/(1 + B))X_1)^{B/(1+B)}, \quad (4.21)$$

with A and B strictly positive integration constants.

Eq. (4.19) states that the value of the elasticity of substitution varies linearly around the intercept of unity and that this value is strictly positive in the 'empirically relevant' range of x (14). Eq. (4.19) can be considered as a first alteration in the basic B.DC.-assumption. Of course, the SATO-REVANKAR class can be easily generalized to homotheticity, with four degrees of freedom W, A, B and b.

From eq. (4.1) we obtain for the MRS-function of the SATO-REVANKAR-class:

$$\hat{s}(x) = B(1 + bx)/x, \text{ domain } \hat{s} = \text{domain } \hat{\sigma}. \quad (4.22)$$

From eq. (4.20) we conclude that the largest possible efficient subsets are not bounded in $\mathbb{R}_{++}^2$ and:

1. If $b \geq 0$, then both boundaries of each largest possible efficient subset do not depend on the parameters of eq. (4.21).
2. If $b < 0$, then only one boundary depend on the parameter b, see also figures 4.1a and 4.1b.

SATO also investigated the ES-function:

$$\sigma(x) = a + bx, \quad (4.23)$$

with:

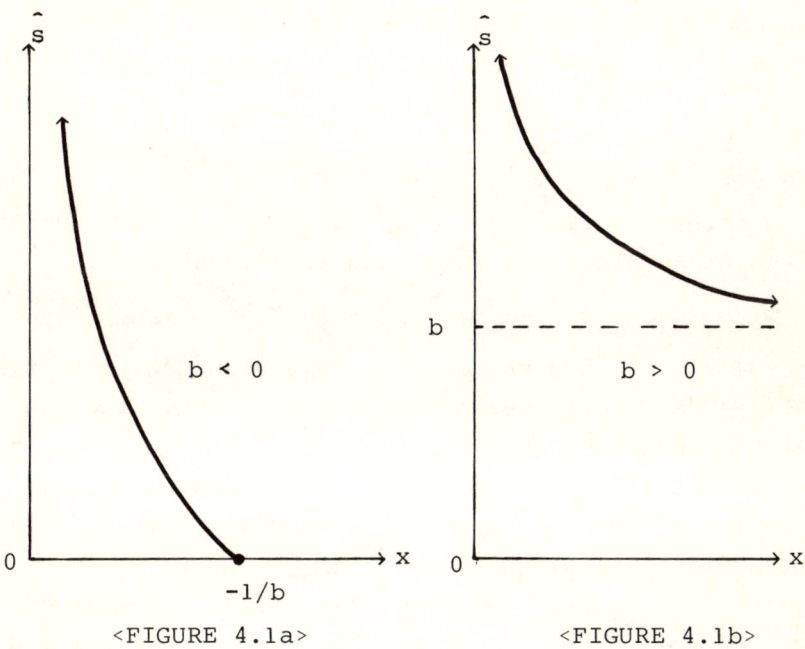

<FIGURE 4.1a>   <FIGURE 4.1b>

$$\text{domain } \hat{\sigma} = \begin{cases} \mathbb{R}_{++}, & \text{if } a \geq 0, b > 0, \\ (0,-b/a), & \text{if } a > 0, b < 0, \\ (-b/a,\infty), & \text{if } a < 0, b > 0. \end{cases} \quad (4.24)$$

SATO remarked that, it is only possible to obtain an explicit form (13) for the implied class of production functions, using eq. (4.3), if in eq. (4.23) the constant a is a rational number (14). It is clear that eq. (4.23) can be considered as a second alteration of the basic B.DC.-assumption, eq. (4.4).

KADIYALA has serious objections against the monotonicity assumption in either eq. (4.4) or eq. (4.23). He states that 'one would expect the elasticity of substitution, $\hat{\sigma}$ to monotonically increase (decrease) as the

input-ratio tends to a "critical" value, say x_o' (15), our symbols. KADIYALA looked for a class of production functions, such that there exists a scalar $x_o \in \mathbb{R}_{++}$ with:
1. Either $\hat{\sigma}' < 0$ if $x < x_o$ and $\hat{\sigma}' > 0$ if $x > x_o$.
2. Or $\hat{\sigma}' > 0$ if $x < x_o$ and $\hat{\sigma}' < 0$ if $x > x_o$.

He suggested as a generalization of the CES-class and the SATO-REVANKAR-class the following functional form:

$$F(X) = (a_{11} X_1^{2\gamma} + 2a_{12} X_1^{\alpha} X_2^{\beta} + a_{22} X_2^{2\gamma})^{\frac{1}{2\gamma}}, \qquad (4.25)$$

with $\alpha + \beta = 2\gamma$, and showed that this class of production functions has the properties mentioned above. Since in KADIYALA's approach the domain of the function $\hat{\sigma}$ equals $\mathbb{R}_{++}$, his class of function can only generate S-neoclassical production functions if the boundaries of the efficient subsets are given ad hoc.

4.3.3. In this subsection we shall discuss three classes of production functions which have been obtained by altering one or more assumptions in the A.C.M.S.-method for obtaining the CES-class, resp. proposed by the A.C.M.S.-group, by BRUNO and by LIU and HILDEBRAND.

We shall first discuss a model which has also been introduced by the A.C.M.S.-group, but which has not been studied by these authors. Instead of the loglinear empirical relation (4.13), they considered (16) the linear relation:

$$Y/X_1 = a + br_1, \quad b > 0, \qquad (4.26)$$

with $Y = F(X)$, F homogeneous of degree one, F unknown.

The assumption $r_1 = F_1'$ and eq. (4.26) yields after rearranging terms, the following differential equation:

$$bX_1 F_1' - F = -aX_1. \qquad (4.27)$$

Eq. (4.27) can be solved by standard-methods. Using the assumption of linear-homogeneity, it yields, either:

$$F(X) = AX_1^{1/b}X_2^{1-(1/b)} + aX_1/(1-b), \quad b \neq 1, \quad (4.28)$$

or:

$$F(X) = (-a)X_1 \ln(A(X_1/X_2)), \quad b = 1 \quad (4.29)$$

with A a strictly positive integration constant. Eq. (4.28) or (4.29) is the functional form of a class of VES production functions. We shall first investigate whether this class of functions can generate S-neoclassical production functions. From eqs. (4.28) and (4.29) we obtain for the RCI-functions of this class:

$$\hat{f}(x) = Ax^{1/b} + \frac{ax}{1-b}, \text{ or } \hat{f}(x) = -ax\ln Ax, \quad (4.30a)$$

$$\hat{f}'(x) = \frac{A}{b}x^{(1/b)-1} + \frac{a}{1-b}, \text{ or } \hat{f}'(x) = -a\ln Ax - a, \quad (4.30b)$$

$$\hat{f}''(x) = A(\frac{1-b}{b^2})x^{(1/b)-2}, \text{ or } \hat{f}''(x) = -a/x. \quad (4.30c)$$

Consider the case $b \neq 1$. From $A > 0$ and $x > 0$ we find that $\hat{f}'' < 0$, if and only if:

$$b > 1. \quad (4.31)$$

From $A > 0$, $x > 0$, $b > 1$ we find that $\hat{f}' > 0$, if and only if:

either $a < 0$, $x > 0$, $\quad (4.32a)$

or $a > 0$, $0 < x < (ab/A(b-1))^{b/(1-b)} := \beta$. $\quad (4.32b)$

From $A > 0$, $x > 0$, $b > 1$ we find that:

$$f - xf' = A(b-1)/bx^{1/b} > 0 \quad (4.33)$$

for all $x \in \mathbb{R}_{++}$.

Since:

$$\hat{f} > \hat{f} - xf' > 0, \quad (4.34)$$

we conclude that the largest possible efficient subsets of eq. (4.28), are generated by either the set:

domain $\hat{f} = \mathbb{R}_+$, $a < 0$, $b > 1$, $\quad (4.35)$

or the set:

$$\text{domain } \hat{f} = [0,\beta], \quad a > 0, \quad b > 1. \tag{4.36}$$

This means that this class of VES production functions has not the property that the largest possible efficient subsets are bounded in $\mathbb{R}^2_{++}$ and either both boundaries, (4.35), or one boundary, (4.36), cannot be written as a function of the parameters of eq. (4.28).

The case $b = 1$ can be studied in an analogous way, the results do not differ from the investigations above.

The class of VES production functions, eqs. (4.28) and (4.29), can be characterized by a simple property. Let the function $\hat{\varepsilon}_1$ of the output elasticity with respect to the input of the first production factor be defined by:

$$\hat{\varepsilon}_1(X) = X_1 F_1'/F. \tag{4.37}$$

Then from:

$$\hat{\varepsilon}_1(x) = x\hat{f}'/\hat{f}(x), \tag{4.38}$$

i.e. ε_1 homogeneous of degree zero, eq. (2.36), system (4.30) and eq. (4.38), it can be easily verified that:

$$\hat{\sigma}(x) = b\hat{\varepsilon}_1(x). \tag{4.39}$$

Eq. (4.39) states that the A.C.M.S.-group proposed a model which is leading to a class of VES production functions, exhibiting the property that the ratio between the functions $\hat{\sigma}$ and $\hat{\varepsilon}_1$ is constant. More generally, it can be shown that, if the function F is homogeneous of degree one, then the functional form of F is given either by eq. (4.28) or by eq. (4.29), if and only if eq. (4.39) holds.

With regard to the empirical relations in the A.C.M.S.-CES-model and the A.C.M.S.-VES-model, BRUNO interprets (17):

1. In the A.C.M.S.-CES-model it follows from eq. (4.13)

that $d\ln(Y/X_1)/d\ln r_1 = \sigma$, F linear-homogeneous, a relative marginal share which is constant.
2. In the A.C.M.S.-VES-model it follows from eq. (4.26) that $d(Y/X_1)/dr_1 = b$, i.e. an absolute marginal share which is constant.

We shall now discuss a functional form which has been obtained by BRUNO (18). This functional form has been specified indirectly on the basis of a model which may be regarded as a modification of the A.C.M.S.-VES-model.

Like in the A.C.M.S.-VES-model, BRUNO assumed that a linear-homogeneous production function $Y = F(X)$ exists and that the linear empirical relation (4.26) holds. He altered the equilibrium condition in the following way:

$$F_1' = c + dr_1. \tag{4.40}$$

BRUNO elucidated his 'disequilibrium and CMS- (= constant marginal shares) model' as follows:
1. 'ARROW, CHENERY, MINHAS and SOLOW have taught us that if one starts from an empirical relation between Y/X_1 and r_1 and assumed equilibrium in the market for the first production factor and the product market, one can work out the implied family of production functions' (in symbols of this study).
2. BRUNO's main concern is to try and use the A.C.M.S.-approach for a situation in which it cannot be assumed that both factor markets are in equilibrium (see eq. (4.40)).
3. According to BRUNO, the linear relation fits better than the logaritmic (CES) version. If $d = 1$, $c = 0$, one gets the competitive A.C.M.S.-VES-model, if $d \neq 1$, $c = 0$, one gets a 'simple case of imperfectly competitive equilibrium' in the first production factor market. The special case $d = 1$, $c < 0$, comes 'close to the empirical situation', if the BRUNO-CMS-model is

applied to data of Israel during the period 1953-1964.

BRUNO remarked that 'the model turns out to give a plausible rationalization of some interesting growth and distribution phenomena which would not make sense on the basis of a more "naive" set of assumptions'.

The BRUNO class of CMS production functions is the result of solving the following differential equation:

$$F = bX_1 F_1'/d + (a - bc/d)X_1, \qquad (4.41)$$

which can be obtained by the elimination of r_1 from eq. (4.26) using eq. (4.40). For $b \neq d$ one gets the following functional form:

$$F(X) = AX_1^{d/b} X_2^{(b-d)/d} + ((ad - bc)/(d - b))X_1. \qquad (4.42)$$

From eqs. (4.28) and (4.42) it follows that the A.C.M.S.-class of production functions is a special subclass of the BRUNO-class of CMS production functions. In an analogous way one can study the largest possible efficient subsets and conclude that these are not bounded in $\mathbb{R}^2_{++}$.

A third example of a functional form specified indirectly on the basis of an alteration in the A.C.M.S.-CES-model has been proposed by LIU and HILDEBRAND (19). They assumed that a linear-homogeneous production function $Y = F(X)$ exists and extended the loglinear empirical relation in the A.C.M.S.-CES-model as follows:

$$\ln Y/X_1 = \ln a + b \ln r_1 + c \ln X_1/X_1. \qquad (4.43)$$

They obtained the following functional form (20):

$$F(X) = (AX_2^{\frac{b-1}{b}} + (a^{\frac{-1}{b}})(\frac{1-b}{1-b-c})(\frac{X_2}{X_1})^{\frac{-c}{b}} X_1^{\frac{b-1}{b}})^{\frac{b}{1-b}}. \qquad (4.44)$$

The properties of this functional form will now be studies along different lines (21). If the first-order conditions for cost minimization under pure competition hold, with r_i = market price i-th production factor, $r_i > 0$, $r := r_1/r_2$, r_i fixed, $i = 1,2$, then the following functional form can be worked out for the implied ES-function:

$$\hat{\sigma}(x) = b/(1 - c(1 + rx)). \tag{4.45}$$

With regard to the sign of the parameter b in eq. (4.43), the following can be said. From EULER's theorem, $F = F_1'X_1 + F_2'X_2$, and from multiplying by $1/X_1$ we obtain:

$$z = \hat{F}_1' + \hat{f}'(u)u, \tag{4.46}$$

where $z := Y/X_1$, $u := 1/x$, $\hat{g}(U) = F(1,u)$. Under pure competition $(r_1 = F_1')$, the result of a disturbance from equilibrium is:

$$dz = dr_1 + (\hat{g}''(u)u + \hat{g}'(u))du. \tag{4.47}$$

From eqs. (4.43) and (4.47) we conclude that, if $du = 0$, $u \neq 0$, then:

$$b = d\ln z/d\ln r_1 = r_1 dz/z dr_1 = r_1/z, \tag{4.48}$$

i.e. the parameter b is strictly positive. This result will now be used for investigating the properties of eq. (4.45). The function $\hat{\sigma}$ is strictly positive and continuous for:

$$0 < x < (1 - c)/cr, \quad 0 < c < 1, \tag{4.49}$$

or

$$x > 0, \quad c < 0. \tag{4.50}$$

In figures 4.2a and 4.2b we illustrated two graphs of the function $\hat{\sigma}$. From (4.45) it follows that the graph of $\hat{\sigma}$ is an orthogonal hyperbole, with the x-axis as a horizontal asymptote. From these figures it can be concluded that eq. (4.44) does not generate largest possible efficient subsets which are bounded in $\mathbb{R}^2_{++}$ and either one

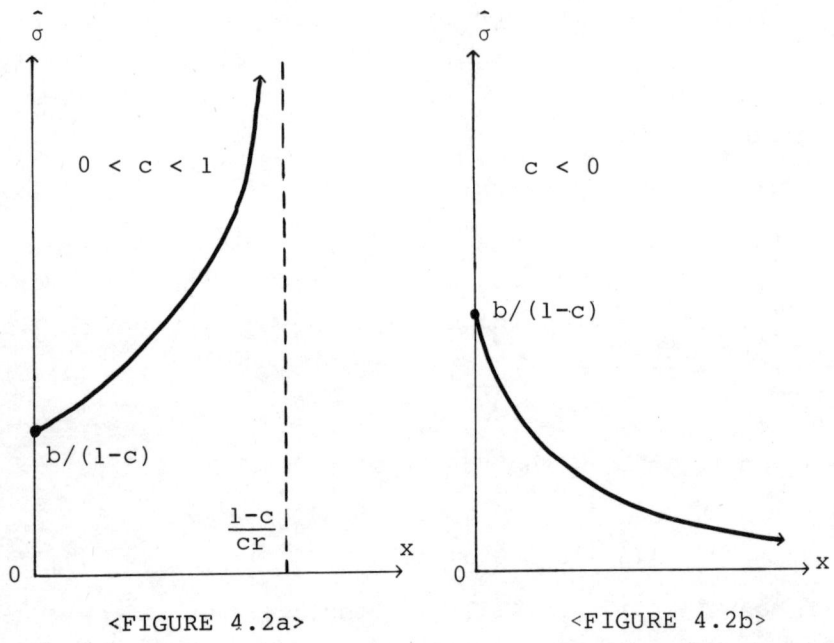

<FIGURE 4.2a>   <FIGURE 4.2b>

boundary, or both boundaries, of such an efficient subset cannot be written as a function of the parameters of eq. (4.44). This means that the largest possible efficient subsets do not fit into the definition of an S-neoclassical production function.

4.3.4. In the last two subsections we discussed several classes of production functions which have been specified in the literature indirectly with the aid of models which are either modifications of the B.DC.-CES-class model, or modifications of the A.C.M.S.-CES-class model. In this subsection we discuss a model of indirectly specifying a class of production functions proposed by SATO and HOFFMANN (22). This model may be regarded as intermediate between the two basic CES-class-generating models.

If it is assumed that F is homogeneous of degree one and both factor markets are competitive, then we may write:

$$\hat{f}'(x)/\hat{f}(x) = r_1 X_1/(FX_1)/X_2, \qquad (4.51)$$

with:

$$r_1 = F'_1. \qquad (4.52)$$

From (4.51) SATO and HOFFMANN obtained:

$$\hat{f}'(x)/\hat{f}(x) = \hat{\alpha}_1(x)/x, \qquad (4.53)$$

where the function $\hat{\alpha}_1(x) := X_1 F'/F$ denotes the function of the income-share of the first production factor. From eq. (4.53) it follows that:

$$\hat{f}(x) = \exp\int \hat{\alpha}_1(x)\,dx/x. \qquad (4.54)$$

If $\hat{\alpha}_1(x)$ is a given function, eq. (4.54) can be used to work out the implied class of production functions. SATO and HOFFMANN proposed the following linear function stating a monotonic relation between the input-ratio and the income-share:

$$\hat{\alpha}_1(x) = px + q, \quad p < 0, \quad 0 < q < 1, \qquad (4.55)$$

and they obtained (23) as the implied class:

$$F(X) = A\exp(pX_1/X_2)X_1^q X_2^{1-q}. \qquad (4.56)$$

Observe that $p = 0$ yields the class of COBB-DOUGLAS production functions as a special subclass of eq. (4.56). The functional form of the related ES-function $\hat{\sigma}$ is:

$$\hat{\sigma}(x) = \frac{-(px + q)(1 - q - px)}{p^2 x^2 + 2pqx + q(q - 1)}. \qquad (4.57)$$

From the conditions imposed on the parameters p and q it follows that the function $\hat{\sigma}$ is strictly positive and continuous for (24):

$$0 < x < -q/p, \quad p \neq 0. \qquad (4.58)$$

This implies that the SATO-HOFFMANN-class, eq. (4.56), does not generate largest possible efficient subsets which are bounded in $\mathbb{R}^2_{++}$. Only one boundary of each largest possible efficient subset depends on the parameters of eq. (4.56), so again a class of functions has been specified that does not fit in the definition of an S-neoclassical production function.

4.4. A CLASS OF VES PRODUCTION FUNCTIONS GENERATING S-NEOCLASSICAL PRODUCTION FUNCTIONS

4.4.1. In this section it is shown that the A.C.M.S.-CES-method can very well be extended, so as to yield a class of production functions for which:
1. The largest possible efficient subsets are bounded in $\mathbb{R}^2_{++}$.
2. The boundaries of each largest possible efficient subset depend on the values of the parameters of the implied functional form.

In subsection 4.4.2. a functional form will be obtained by altering two basic assumptions in the A.C.M.S.-CES-model. For the case of constant returns to scale this functional form has been independently obtained by BRUNO and by VAZQUEZ (25). In subsection 4.4.3. it will be shown that this functional form actually generates largest possible efficient subsets which are bounded in $\mathbb{R}^2_{++}$, the boundaries depending on the values of the parameters of the implied functional form.

Subsection 4.4.4. consists of a table in which the assumptions underlying the various functional forms, discussed in this chapter, are summarized.

4.4.2. Let $Y = F(X)$ be an unknown two-factor production function. It will be assumed that the function F is homogeneous of degree $\lambda > 0$. Let the variable Z be defined by:

$$Z := Y^{1/\lambda}, \qquad (4.59)$$

or in an equivalent way by:

$$Z := \hat{F}(X), \qquad (4.60)$$

where $\hat{F}(X) = (F(X))^{1/\lambda}$. The variable Z may be interpreted as an index of the inputs of the production factors (29), or as the maximum output rate of the input-combination (X_1, X_2), if constant returns to scale would prevail.

Like in the A.C.M.S.-CES-model it will be assumed that the output-price is fixed at the unit level and that the price for the use of one unit of the first production factor is given.

The empirical loglinear relation in the A.C.M.S.-CES-model is altered in the following way:

$$Z/X_1 = a + br_1 + cX_2/X_1, \qquad (4.61)$$

i.e. it is assumed that a linear relation exists between average constant returns to scale productivity, the price of the first production factor and the reciproce of the input-ratio.

Instead of the equilibrium condition in the A.C.M.S.-model, it will be assumed that:

$$r_1 = (dZ/dY) F_1'(X). \qquad (4.62)$$

This assumption implies that:
1. If F is homogeneous of degree one, then the first production factor marginal productivity equals the first production factor price.
2. Normalizing the variable Z to unity, i.e. $\hat{F}(X) = 1$, yields that (i) decreasing returns to scale leads to an excess of the first factor price over its marginal productivity, and (ii) increasing returns to scale leads to an excess of the first factor marginal productivity over its price, since:

$$dZ/dY = \frac{1}{\lambda} Z^{(1-\lambda)/\lambda}. \qquad (4.63)$$

Eq. (4.62) may be regarded as a modification of the disequilibrium assumption in the BRUNO-CMS-model. Its implications can also be elucidated in the following way. The multiplication of eq. (4.62) by X_1, yields:

$$r_1 X_1 = (YdZ/ZdY)(F_1' X_1/Y) Z, \qquad (4.64)$$

or:

$$r_1 X_1 = (\varepsilon_1(x)/\lambda) Z. \qquad (4.65)$$

Eq. (4.65) states that total rewards of the first production factor are proportionate to the (real) value of the constant returns to scale output, where the propertionality factor is a function of the degree of homogeneity and the output-elasticity of the input of one unit of the first production factor.

It will now be shown that this extended BRUNO-VAZQUEZ-model, eqs. (4.59), (4.61), (4.62), can be used for the indirect specification of a functional form of a class of production functions.

Using the chain-rule for differentiating a compound function results into:

$$\partial Z/\partial X_1 = (dZ/dY) F_1', \qquad (4.66)$$

or from eq. (4.62):

$$r_1 = \hat{F}_1'. \qquad (4.67)$$

By the elimination of r_1 from eq. (4.61), using (4.67), the following differential equation in the unknown function $\hat{F}$ is obtained:

$$\hat{F} = aX_1 + bX_1 \hat{F}_1' + cX_2. \qquad (4.68)$$

This differential equation can be solved by standard-methods, from which we easily obtain:

$$\hat{F}(X) = AX_1^{1/b} X_2^{1-1/b} + aX_1/(1-b) + cX_2, \quad (4.69)$$

for $0 \neq b \neq 1$, or:

$$\hat{F}(X) = -aX_1 \ln(X_1/X_2) + cX_2 + AX_1 \quad (4.70)$$

for $b = 1$, with A a strictly positive integration constant. That is to say, for the function F the following functional form has been specified indirectly:

$$F(X) = (AX_1^{1/b} X_2^{1-1/b} + aX_1/(1-b) + cX_2)^\lambda, \quad (4.71)$$

or:

$$F(X) = (-aX_1 \ln(X_1/X_2) + cX_2)^\lambda. \quad (4.72)$$

4.4.3. It will now be proved that the BRUNO-VAZQUEZ-class of production functions, eq. (4.71), has the following properties:
1. This class of functions generates largest possible efficient subsets which are bounded in $\mathbb{R}_{++}^2$.
2. The boundaries of each largest possible efficient subset depend on the parameters of this class.

Assume that $0 < b \neq 1$ and consider the RCI-function of eq. (4.71). For the function $\hat{f}$ we obtain:

$$\hat{f}(x) = Ax^{1/b} + ax/(1-b) + c, \quad (4.73a)$$

$$\hat{f}'(x) = (A/b)x^{(1-b)/b} + a/(1-b), \quad (4.73b)$$

$$\hat{f}''(x) = ((1-b)A/b^2)x^{(1-2b)/b}. \quad (4.73c)$$

From $A > 0$ and eq. (4.73c) we conclude that $\hat{f}'' < 0$, if and only if:

$$b > 1. \quad (4.74)$$

From $A > 0$, eq. (4.73b) and ineq. (4.74) it follows that $\hat{f}' > 0$, either for all $x \in \mathbb{R}_{++}$ if $a \leqq 0$, or for:

$$x^{(1-b)/b} > ab/A(b-1) := q, \quad a > 0. \quad (4.75)$$

From $A > 0$, eqs. (4.73a), (4.73b) and ineq. (4.74) we

obtain that:

$$\hat{f} - x\hat{f}' = (A(b-1)/b)x^{1/b} + c > 0, \quad (4.76)$$

either for all $x \in \mathbb{R}_{++}$ if $c \gtreqless 0$, or:

$$x^{1/b} > bc/A(1-b) := p, \; c < 0. \quad (4.77)$$

Using $\hat{f} > \hat{f} - x\hat{f}' > 0$, ineqs. (4.74), (4.75), (4.77), yields the following proposition. The functional form (4.73a) is a class of RCI-functions with:

i) largest possible efficient subsets which are bounded in $\mathbb{R}^2_{++}$ and

ii) both boundaries of each largest possible efficient subset depend on the value of the parameters of this class,

if and only if:

i) $A > 0, \; a > 0, \; b > 1, \; c < 0$,

then the largest possible domain is defined by:

ii) $0 < \alpha < \beta < \infty, \; \alpha \leq x \leq \beta, \; \alpha := p^b, \; \beta := q^{b/(1-b)}$.

This means that, if the inequalities in this theorem hold, then eq. (4.71) defines a class of RP-functions, and the extended BRUNO-VAZQUEZ-model specifies indirectly a class of S-neoclassical production functions.

It is clear that the border-case $b = 1$ can be studied in an analogous way.

4.4.4. In this subsection we given an outline of the assumptions and the properties of the various functional forms which have been discussed. In part A of the table below we summarize the B.DC.-CES-method and its extensions, in part B the A.C.M.S.-CES-method and its extensions. The SATO-HOFFMANN-model is considered as an extension of the A.C.M.S.-model.

TABLE: Survey of indirectly specified classes of neoclassical production functions

A. AUTHOR(S)	Brown–De Cani	Sato/Revankar	Sato		Bruno/Vazquez
Technical assumptions	homothetic $\hat{\sigma}$ constant	homothetic $\hat{\sigma} = 1 + bx$	homothetic $\hat{\sigma} = a + bx$		homogeneous
Functional form	(4.10),(4.11)	(4.21)	Exists if a is rational		$r_1 = Z'(Y)F_1'$ $Z/X_1 = a+br_1 + cX_2/X_1$
Larg.poss. dom.RCI-func.	$\mathbb{R}_+$	$\mathbb{R}_+, [0,-1/b]$	$\mathbb{R}_+, [0,-b/a],$ $[-b/a,\infty)$		(4.71),(4.72) $[\alpha,\beta]$

B. AUTHOR(S)	A.C.M.S./ Paroush	A.C.M.S.	Bruno	Liu-Hildebrand	Sato-Hoffmann	Bruno/Vazquez
Technical assumption	homogeneous	homogeneous degree one	homogeneous degree one	homogeneous degree one	homogeneous degree one	homogeneous
Economic assumption	$r_1 = F_1'$	$r_1 = F_1'$	$F_1' = c + dr_1$	$r_1 = F_1'$	$r_1 = F_1'$	
Empirical relation	$\ln Y = \ln a +$ $b\ln r_1 + c\ln X_1$	$Y/X_1 = a+br_1$	$Y/X_1 = a+br_1$	$\ln Y/X_1 = \ln a +$ $b\ln r_1 + c\ln X_2/X_1$	$\hat{\alpha}_1 = px+q$	
Functional form	(4.14),(4.15)	(4.28),(4.29)	(4.42)	(4.44)	(4.56)	
Larg.poss. dom.RCI-func.	$\mathbb{R}_+$	$\mathbb{R}_+, [0,\beta]$	$\mathbb{R}_+, [0,\beta]$	$\mathbb{R}_+, [0, \frac{1-c}{cr}]$	$[0,-q/p]$	

NOTES

1. For recent uses, see for example EICHHORN, HENN, OPITZ, SHEPHARD (1974).
2. DIEWERT (1976).
3. ARROW, CHENERY, MINHAS, SOLOW (1961); BROWN, De CANI (1963). See also HARDY, LITTLEWOOD, PÓLYA (1934).
4. For a discussion see De JONG, KRISHNA KAMUR (1972).
5. For alternative derivations, related to the BROWN-De CANI-method, see VAZQUEZ (1971).
6. SATO (1967).
7. For a discussion see BROWN (1968), p. 44. In chapter 2 of this study, we defined several 'second-stage' functions, like the RP-, RCI-, ES-function, in terms of some given S-neoclassical production function. It is clear that all these second-stage functions can in an analogous way be defined in terms of some given traditional neoclassical production functions, see VAZQUEZ (1971), FØRSUND (1974).
8. See PAROUSH (1964); YASUI (1965).
9. ALLEN (1972), page 54; De JONG, KRISHNA KAMUR (1972).
10. Linear-homogeneity: ARROW, CHENERY, MINHAS, SOLOW (1961); homogeneity: PAROUSH (1966).
11. For a derivation, see VAZQUEZ (1961).
12. SATO (1967), REVANKAR (1971a), (1971b).
13. Whether an explicit functional form exists is from a theoretical point-of-view not important. In our opinion the emphasis must be: the existence of some class of production functions and its implied properties.
14. SATO (1967). We believe that in SATO's theorem $b \neq 0$. The direct application of this theorem would lead to the incorrect conclusion that the CES-class is only defined for rational numbers.
15. KADIYALA (1972). DENNY (1974) argues, 'that there is little justification for making the elasticity of substitution a monotone function of the input-ratio'.
16. ARROW, CHENERY, MINHAS, SOLOW (1961).
17. BRUNO (1968).
18. BRUNO (1968).

19. LIU, HILDEBRAND (1965); see also LU (1967) and LU and FLETCHER (1968).

20. De BOER (1971) showed that the SATO-REVANKAR-class, eq. (4.21), may be considered as a subclass of the LIU-HILDEBRAND-class, eq. (4.42). De BOER incorrectly called the LIU-HILDEBRAND-class the SATO-HOFFMANN-class. The so-called SATO-HOFFMANN-class will be discussed in subsection 4.3.4.

21. LU, FLETCHER (1968), p. 450. FLECK, FINKBEINER, CASUTT (1971), p. 68.
    For critisism and discussions regarding the LIU-HILDEBRAND-class see FLECK, FINKBEINER, CASUTT (1971), BARTNICK, GAHLEN (1972), LAüFER (1974).

22. SATO, HOFFMANN (1968).

23. KNOX LOVELL (1973) noted that eq. (4.56) has also been proposed by FERGUSON. Eq. (4.56) has been generalized into two ways. First:
$$\hat{f}(x) = A\exp\int \hat{\alpha}_1(x)dx/x = A\exp \Sigma\, a_k x^k.$$
so-called polynomial curve-fitting, CLEMHOUT (1968). Secondly:
$$Y = A(aX_1^{-b} + (1-a)X_2^{-b})^{-1/b} \exp(cx^d),$$
a hybrid of a CES-function and an exponented power function, see De BIOLLEY (1972).

24. The nominator of eq. (4.57) has one strictly positive root $x = -q/p > 0$. The denominator has one strictly positive root $x = (-q - \sqrt{q})/p > -q > p$. This implies that the ratio in eq. (4.57) will be positive for $0 < x < -q/p$.

25. VAZQUEZ (1971), footnote 35, quoting BRUNO (1962), unpublished memorandum.

5. The direct specification of classes of neoclassical production functions

5.1. INTRODUCTION

5.1.1. In this chapter we investigate the properties of a number of functional forms which have been specified directly in order to represent classes of production functions.

We say that a functional form is *specified directly*, if this functional form is introduced heuristically, i.e. not as in chapter 4 the result of integrating a certain differential equation, which itself is based on assumptions regarding partial derivatives of an unknown production function. The direct specification may be regarded as a manner to perform a very important step in e.g. the evaluating of production functions. In this chapter the following problems will be studied:
1. On what arguments can the direct specification of a functional form be based?
2. Has such a functional form the properties, that it can be used in representing S-neoclassical production functions and does it generate largest possible efficient subsets which are bounded in $\mathbb{R}^2_{++}$? Do the boundaries of each largest possible efficient depend on the parameters of the functional form?

5.1.2. Traditional empirical methods are mostly based on simple, directly specified, functional forms, e.g. linear or loglinear ones. HANOCH believes that several concurrent developments increased the interest in more flexible and more general functional forms to be used in empirical applications of production functions. As such developments HANOCH mentions, improvements in econometric methods and in computer capacity, more detailed data and increased interest in new types of problems (1). HANOCH states that (2), if the *direct specification* of a functional form is based on the *local approximation approach*, then this approach has clear advantages for the case that the number of variables is moderately small, say two to five. In such a case the number of parameters is relatively small too. From the characteristics of our assumed technology (see chapter 2), it follows that we actually have to deal with a small number of variables: one dependent variable Y and two independent variables X_1 and X_2. Moreover, we believe that the approximation approach is more in line with an empirical point of view, see also the next section.

As will be argued in section 3.3., the direct specification of a functional form, based on the local approximation approach, can be *decompounded into two independent steps, local approximation of the RCI-function or local approximation of the RC-function (substitutability) and local approximation of the transform (scale properties)*. This decompounding of a local approximation is of basic importance for the discussions in this chapter.

5.1.3. In section 5.2. we discuss the concept of approximation of a given function, in section 5.3. the concept of local approximation of a given function. In section 5.4. we investigate the properties of functional forms proposed by KUIPERS, DIEWERT, BRUNO/VAZQUEZ and DENNY. In section 5.5. we pay attention to functional forms proposed by ZELLNER/REVANKAR and RINGSTAD and a general class of so-called FRISCH transforms. This chapter concludes with section 5.6, in which a number of concluding remarks are made and proofs of two theorems are presented.

5.2. GENERAL REMARKS ON THE CONCEPT OF APPROXIMATION

5.2.1. In several recent contributions (3) a number of functional forms are introduced (specified) directly and their properties investigated from an approximation-theoretical point of view. We shall first try to answer the questions *why* the concept of approximation is used and *how* this concept is applied in these studies. We believe that in the contributions referred to, the answers presented are not clear enough.

First, one must conclude (4) that in an approximation-approach the *motive* for directly specifying a functional forms is bearing an economic-*empirical* character, since in an empirical analysis mostly 1) it is assumed that a (production) function exists with properties to be postulated, 2) actually this function is unknown, 3) in a direct way a functional form is defined in which no numerical values are attributed to several parameters, 4) from the use of empirical observations one tries to obtain a combination of values for these parameters, 5) the function obtained in step 4 is considered as an approximation of the unknown function in step 1. That is to say, in an empirical analysis one has to replace each function, which enters a theoretical model, by a functional form. This step must precede the last step: empirical estimation and testing.

The *way* of discussing approximation in the literature must be described in a different manner, i.e. a *theoretical* one, since in approximation-theoretical studies 1) it is assumed that a function is given with all parameters numerically specified, 2) a functional form is specified directly in order to represent a class of functions (this class we call the *approximating* class), 3) from this class a unique function is selected, 4) the function, selected in step 3, is considered as an approximation of the given function in step 1.

In the contributions referred to, approximation is discussed in a theoretical manner, the underlying idea is that an approximating class can be used for an analysis in an empirical sense.

In this chapter the direct specification of functional forms is based on the approximation approach. That is to say, in our discussions we follow the approximation-theoretical approach: it will be assumed that we want to approximate a given function by using a directly specified functional form, which represents an approximating class of functions.

5.2.2. In the application of the concept of approximation one must distinguish two separate issues, 1) whether it is desirable to approximate a given function *globally*, or 2) whether it is desirable to approximate a given function *locally*. LAU believes (5) that for certain types of economic problems (e.g. dynamic ones) it may be desirable to approximate globally, but in most applied econometric studies local approximation will suffice. He argues, 'it does not make too much sense to extrapolate the results too far from the sample anyhow' (6).

In this study we take the local approximation approach as a starting-point. This means that it will be assumed *that an S-neoclassical production function* H *is given and that the function* H *is approximated locally in a small neighbourhood of a fixed point* X $\in$ *interior cone* $\overline{D}$, *by the selection of a function* G *from an approximating class of functions, where the class of functions is defined by means of a directly specified functional form.*

5.3. INTRODUCTORY REMARKS ON LOCAL APPROXIMATION

5.3.1. It must be noted that a local approximation of a given function, which itself shows a number of properties, does not preserve these properties in general. LAU observes (7) that 'care should be exercized in making inferences from the properties of a true function on the basis of the approximating function'. Furthermore, if an approximating class of functions is given, then one can investigate for which values of the parameters and independent variables, each function from this class will be an S-neoclassical production function. From a mathematical point of view such an investigation is superfluously, since the properties of an approximating function hold only locally with respect to a given function. From an empirical point of view such an investigation is, however, not superfluously. For example constraints on parameters of a functional form may be the basis of tests of the validity of this functional form (8).

In order to be able:
a. to get a better understanding of recent discussions on local approximation (see note 3 this chapter) and
b. to apply the concept of local approximation to the concept of S-neoclassical production functions,
c. to link the local approximation approach with the empirical point of view, that a functional form can be used for testing the assumption of bounded efficient subsets,

we shall explicitly impose an additional restriction with regard to the approximating function G of a given S-neoclassical production function H. It will be imposed that *the approximating function* G *itself must be an S-neoclassical production function*. In the next sections we are looking for functional forms which (i) will be specified

directly and (ii) will generate production functions
which on their largest possible domain are S-neoclassical.

5.3.2. From the investigations in chapter 2 it follows
that an S-neoclassical production function H is a
compound function of a transform W applied to a core
function $\hat{H}$. This means that for an S-neoclassical production function the *scale* properties and the *substitution*
properties can be studied independently (9). Or, if a
given S-neoclassical production function H is approximated locally with respect to some point $X^o \in$ int.
cone $\overline{D} := D$, then:

1. The substitution properties can be approximated locally by means of an approximation $\hat{U}$ of the RC-function $\hat{F}$
   in X^o, or an approximation $\hat{u}$ of the RCI-function $\hat{f}$ in
   $x^o := X_1^o/X_2^o$.
2. The scale properties can be approximated locally by
   means of an approximation V of the transform W in
   $z^o := \hat{F}(X^o)$.

It can be easily proved that an approximation of an
S-neoclassical production function H itself, by means of
an n-th order TAYLOR expansion, cannot be written as a
compound function of a transform and an RC-function.
This means that a TAYLOR-approximation cannot represent
an S-neoclassical production function (see also the discussion in subsection 5.4.1.). The restriction that a
local approximation G of the function H must itself be an
S-neoclassical production function leads to the conclusion that the function G must be a compound function of
an approximation of W and an approximation of $\hat{F}$. In section 5.4. we discuss functional forms which perform a
local approximation of a given RC-function $\hat{F}$ or a given
RCI-function $\hat{f}$ (the substitution properties), in section

5.5. we discuss functional forms which can approximate locally a given transform W (the scale properties), such that:
1. The compound function of both local approximations, say $V(\hat{U}(X))$, is an RP-function of an S-neoclassical production function.
2. The compound function $V(\vec{U}(X))$ can be extended to an S-neoclassical production function by means of (2.11c), (2.11d) and (2.11e).

5.3.3. The mathematical basis of decompounding a local approximation into two independent steps consists of two theorems, one definition and one lemma. This mathematical basis will now be given in terms of functions with *two* independent variables, but can be generalized to functions of m independent variables. The proofs of the two theorems below will be given in section 5.6. of this chapter.

THEOREM 5.1: Let the real-valued functions $u(X)$ and $v(X)$, $X \in \mathbb{R}^2$, be n times continuously differentiable on a non-empty open ball $B_r(X^o)$.
Denote the partial derivatives of arbitrary functions $w(X_1, X_2)$ by:
$$w_{p,k} := \partial^p w / \partial X_1^k \partial X_2^{p-k} \qquad (5.1)$$
Then with LANDAU's symbol o(.), (10)
$$u(X^o+h) - v(X^o+h) = o(\|h\|^n), \|h\| \to 0, h \in \mathbb{R}^2 \quad (5.2)$$
if and only if:
$$u(X^o) = v(X^o), (u_{p,k} = v_{p,k})_{X=X^o}, \qquad (5.3)$$
for all pairs (p,k) satisfying $1 \leq p \leq n$, $0 \leq k \leq p$.

This theorem states that in the neighbourhood of X^o the difference between the two functions u and v is an infinitesimal of order higher than $\|h\|^n$, if and only if the values of u and v and all their partial derivatives up to and including those of the n-th order, are equal for $X = X^o$.

DEFINITION 5.1: The real-valued function $v(X)$, $X \in \mathbb{R}^2$, is a *local n-th order approximation of* the real-valued function $u(X)$ at the point X^o, if and only if in a neighbourhood of X^o contained in domain $u(X) \cap$ domain $v(X)$:

$$u(X^o+h) - v(X^o+h) = o(\|h\|^n), \quad h \in \mathbb{R}^2, \quad \|h\| \to 0. \quad (5.4)$$

Theorem 5.1 and Definition 5.1 can also be formulated for functions u and v with either one independent variable or more than two independent variables. For the case of one independent variable this is summarized in lemma 5.1. Note that an n-th order approximation (n > 1) is also an (n - 1)-th order approximation.

LEMMA 5.1: The real-valued n times continuously differentiable function $g(x)$, $x \in \mathbb{R}$, is a local n-th order approximation of the real-valued n times continuously differentiable function $f(x)$, if and only if:

$$f(x^o) = g(x^o), \quad f^{(p)}(x^o) = g^{(p)}(x^o), \quad (5.5)$$

for $p = 1,\ldots,n$.

That a compound approximation, as introduced in subsection 5.3.2. above, can perform a local approximation of a given function H follows from the following theorem.

THEOREM 5.2: Let the real-valued functions $u(X)$ and $v(X)$, $X \in \mathbb{R}^2$, be n times continuously differentiable on a neighbourhood of $X^o \in \mathbb{R}^2$ and let the function $v(X)$ be a local n-th order approximation of the function $u(X)$ in X^o. Let the real-valued functions $f(x)$ and $g(x)$, $x \in \mathbb{R}$, be n times continuously differentiable on a neighbourhood of $x^o \in \mathbb{R}$ and let the function $g(x)$ be a local n-th

order approximation of the function f(x) in
$x^o = u(X^o) \in$ domain f $\cap$ domain g. Then the compound function g(v(X)) is a local n-th order approximation of the compound function f(u(X)) in X^o.

5.4. LOCAL APPROXIMATION OF RC- AND RCI-FUNCTIONS

5.4.1. For a given S-neoclassical production function H
it follows that the RCI-function $\hat{F}$ or the RCI-function $\hat{f}$ contains all information about technical substitutability. This can be concluded from the property that the MRS-function $\hat{s}$ and the ES-function $\hat{\sigma}$ of a given S-neoclassical production function H can be expressed in terms of X_1, X_2, $\hat{F}$, $\hat{F}'_1$, $\hat{F}'_{11}$, or of x, $\hat{f}$, $\hat{f}'$, $\hat{f}''$ (see eqs. (2.31) and (2.36)).

Let $\hat{F}(X)$ be the RC-function of a given S-neoclassical production function H and let $X^o \in D$. In this subsection we shall discuss TAYLOR-approximations of $\hat{F}$. First, if we approximate the function $\hat{F}$ by means of the class of linear functions:

$$\hat{U}(X) := a_{00} + a_{01}X_1 + a_{02}X_2, \tag{5.6}$$

i.e. a first-order TAYLOR-approximation, or, secondly, by means of the class of functions:

$$\hat{U}(X) := a_{00} + a_{01}X_1 + a_{02}X_2 + a_{11}X_1^2 + a_{12}X_1X_2 + a_{22}X_2^2, \tag{5.7}$$

a second-order TAYLOR-approximation, then we conclude:
1. From eq. (5.6), that $\hat{U}''_{ij} = 0$, i,j = 1,2, for all X.
2. From eq. (5.7) and the linear-homogeneity of $\hat{F}$, that each $\hat{U}$ is an RC-function if and only if $a_{00} = a_{11} = a_{12} = a_{22} = 0$.

It can be easily proved that, if, the approximation $\hat{U}$ must be an RC-function, then any n-th order TAYLOR-approximation does not represent a class of RC-functions, since each n-th order TAYLOR-approximation is only homogeneous of degree one, if this approximation reduces to a function of the form of eq. (5.6) with $a_{00} = 0$.

5.4.2. Let us now consider the RCI-function $\hat{f}$ of a given S-neoclassical production function H. It is obvious to try subsequently a TAYLOR-approximation of the given function $\hat{f}$. If we first approximate the function $\hat{f}^o$ in $x^o \in (a,b)$, by means of a first-order TAYLOR-expansion:

$$\hat{u}(x) := a_0 + a_1 x, \qquad (5.8)$$

then we obtain $\hat{u}'' = 0$. So a first-order TAYLOR-approximation cannot be used for a local approximation of the function $\hat{f}$, since the class of functions defined by eq. (5.8) is not flexible enough to represent RCI-functions.

Secondly, let the class of functions $\hat{u}$ be defined by:

$$\hat{u}(x) := a_0 + a_1 x + a_2 x^2, \qquad (5.9)$$

i.e. we try to approximate $\hat{f}$ by means of a second-order TAYLOR-expansion. We obtain a unique combination of values for the parameters a_0, a_1 and a_3 such that (11):

$$\hat{u}^o = a_0 + a_1(x^o) + a_2(x^o)^2 = \hat{f}^o, \qquad (5.10a)$$

$$(\hat{u}')^o = a_1 + 2a_2 x^o = (\hat{f}')^o, \qquad (5.10b)$$

$$(\hat{u}'')^o = 2a_2 = (\hat{f}'')^o. \qquad (5.10c)$$

From system (5.10) and lemma 5.1 it follows that eq. (5.9) defines a class of functions which can indeed perform a local second-order approximation of the given function $\hat{f}$ in each $x^o \in (a,b)$.

Let us now investigate whether eq. (5.9) defines a class of RCI-functions whose largest possible domains are bounded in $\mathbb{R}_{++}$ and for which the boundaries of each largest possible domain depend on the parameters of

(5.9). If the answers to these two questions are in the affirmative, then (5.8) can be used in an empirical testing of SHEPHARD's boundedness assumption.

We conclude that $\hat{u}'' < 0$, if and only if:

$$a_2 < 0. \tag{5.11}$$

From (5.11) it follows that $\hat{u}' > 0$, if:

$$x^o < -a_1/2a_2 := \bar{b}, \ a_1 > 0. \tag{5.12}$$

From (5.11) we find that $\hat{u} - x\hat{u}' > 0$, if:

$$x^o > (a_0/a_2)^{\frac{1}{2}} := \underline{a}, \ a_0 < 0. \tag{5.13}$$

From (5.12) and (5.13) we obtain that $\underline{a} < \bar{b}$, if and only if:

$$a_1^2 - 4a_0a_2 > 0. \tag{5.14}$$

From (5.11) to (5.14) we conclude that the directly specified class of functions (4.9), generates a class of RCI-functions whose largest possible domains are bounded in $\mathbb{R}_{++}$ and for which each largest possible domain has boundaries which depend on the parameters, if and only if:

$$a_0 < 0, \ a_1 > 0, \ a_2 < 0, \ a_1^2 - 4a_0a_2 > 0, \tag{5.15}$$

and $[\underline{a}, \bar{b}] \subset \mathbb{R}_{++}$ is the largest possible domain of each function $\hat{u}$.

If the inequalities in (5.15) hold, then eq. (5.9) can be used to define the following class of RC-functions:

$$\hat{U}(X) := X_2\hat{u}(X_1/X_2) = a_0X_2 + a_1X_1 + a_2X_1^2/X_2. \tag{5.16}$$

It can be easily proved that eq. (5.26) can be used to accomplish a local second-order approximation of a given RC-function $\hat{F}$. Namely, from $\hat{U}(X_1,X_2) = X_2\hat{u}(X_1/X_2)$, $\hat{F}(X_1,X_2) = X_2\hat{f}(X_1/X_2)$, theorem 5.1., definition 5.1 and lemma 5.1., we conclude that $\hat{U}$ is a local approximation of $\hat{F}$, if and only if $\hat{u}$ is a local approximation of $\hat{f}$. Eq.

(5.26) can, however, *not* be used to accomplish a second-order TAYLOR-approximation of a given RC-function $\hat{F}$.

The functional form (5.16) happens to be introduced by KUIPERS in a direct way, as a class of linear-homogeneous production functions (12). KUIPERS (i) did not investigate completely whether his class of production functions fits into a view of production technologies and (ii) did not explicitly use the concept of local approximation. He used this class as an ingredient in a one-sector growth model and estimated the values of the parameters a_0, a_1 and a_2, applying empirical observations of the Dutch economy. One of the explicit purposes of KUIPERS is to give a description of a technology which exhibits the property of *bounded substitutability* (13), in SHEPHARD's terminology bounded efficient subsets. The directly specified class of KUIPERS may be considered as a precursor of a number of directly specified classes of functions which have been based on the local approximation approach.

It must be noted that for the KUIPERS-class, the definition of the input-ratio is crucial. With respect to the variable $x = X_1/X_2$, eq. (5.9) can be the basis for a TAYLOR-approximation of a given RCI-function $\hat{f}$. With respect to the variable $1/x := X_2/X_1$, eq. (5.9) cannot be the basis for a TAYLOR-approximation of a given RCI-function $\hat{f}$.

This type of 'asymmetry' (14) between the variable x and $1/x$, does not occur in a functional form proposed by DIEWERT (15). DIEWERT considered the following functional form:

$$\hat{U}(X) := a_1 X_1 + a_2 (X_1 X_2)^{\frac{1}{2}} + a_3 X_2. \tag{5.17}$$

From (5.16) we find:

$$\hat{u}(x) := \hat{U}(x,1) = a_1(\sqrt{x})^2 + a_2\sqrt{x} + a_3 \qquad (5.18)$$

and

$$\hat{v}(1/x) := \hat{U}(1,1/x) =$$
$$= a_1 + a_2\sqrt{(1/x)} + a_3(\sqrt{(1/x)})^2. \qquad (5.19)$$

This directly specified class of functions $\hat{U}$ can perform a local second-order approximation, not a TAYLOR-approximation, of a given RC-function $\hat{F}$. This follows from the fact that, if it is assumed that $\hat{U}^o = \hat{F}^o$, $(\hat{U}')^o = (\hat{F}')^o$ and $(\hat{U}'')^o = (\hat{F}'')^o$, then eq. (5.17) can be used for the construction of a system of equations, which yields a unique combination of values for the parameters of eq. (5.17), (16). The class of functions $\hat{u}$ defined by eq. (5.18) can accomplish a second-order TAYLOR-approximation of a given function $\hat{f}$ in terms of the variable $x^{\frac{1}{2}}$. The class of function $\hat{v}$ defined by eq. (5.19) can accomplish a second-order TAYLOR-approximation of a given function $\hat{f}$ in terms of the variable $(1/x)^{\frac{1}{2}}$.

Like for the KUIPERS class of functions, it can be proved that, eq. (5.18) defines a class of RCI-functions whose largest possible domains are bounded in $\mathbb{R}_{++}$ and for which each largest possible domain has boundaries which depend on the parameters of eq. (5.18), if and only if:

$$a_1 < 0, \; a_2 > 0, \; a_3 < 0, \; a_2^2 - 4a_1 a_3 > 0, \qquad (5.20a)$$
$$0 < 4a_3^2/a_2^2 := \underline{a} \leq x = \overline{b} :\leq a_2^2/4a_1^2 < \infty. \qquad (5.20b)$$

If the inequalities in (5.20a) hold, then eq. (5.16) defines a class of RC-function $\hat{U}$ generating largest possible efficient subsets which are bounded in $\mathbb{R}_{++}^2$.

In this subsection we investigated the properties of two functional forms proposed by KUIPERS and by DIEWERT. Our results may be summarized as follows:

1. Both functional forms (see resp. (5.16), (5.17)), can perform a local second-order approximation of a given RC-function $\hat{F}$.
2. For certain values of their parameters and independent variables, both functional forms define classes of RC-functions whose largest possible efficient subsets are bounded in $\mathbb{R}^2_{++}$ and for which each largest possible efficient subset has boundaries which depend on the parameters.
3. The KUIPER's class of RCI-functions $\hat{u}$ (see eq. (5.19)) can perform a TAYLOR-approximation of a given RCI-function $\hat{f}$, but is 'asymmetric'.
4. The DIEWERT-class of RCI-functions $\hat{u}$ (see eq. (5.18) can perform a TAYLOR-approximation of a given RCI-function $\hat{f}$ in terms of the variable $\sqrt{x}$, and is 'symmetric', i.e. the definition of the input-ratio is not crucial.

5.4.3. In the preceding subsection we discussed two functional forms which could perform a local second-order approximation of a given RC-function (17). LAU states (18) that 'If more flexibility is needed, one can extend (...) functions to the third order in a straight forward manner. In general, however, second order numerical approximations seem quite adequate. And since all second order numerical approximations are equivalent, one can choose the one functional form that is most suitable for the problem at hand'. There can, however, be several reasons to prefer approximations of an order higher than the second. E.g.:
1. The technology may be defined in such a way that a local second order approximation does not suffice. LAU points to a study of FUSS and McFADDEN concerning the two-stage (ex ante, ex post) description of a production technology (19).

2. Production functions are not isolated concepts in economic analyses. Mostly a production function is one of the main ingredients of a quantitative model. The purpose of the construction of a model may determine the order of approximation of a production function. E.g., in several recent contributions (20) qualitative properties are attributed to the behavior of the ES-function $\hat{\sigma}$ of a neoclassical production function, for example that the function $\hat{\sigma}$ is strictly monotone increasing, i.e. $\hat{\sigma}$ differentiable and $\hat{\sigma}' > 0$. If values for x, $\hat{f}$, $\hat{f}'$ and $\hat{f}''$ are given, then one can only obtain the value for $\hat{\sigma}$ (see eq. (2.36)). If one wants to obtain a value for $\hat{\sigma}'$, this value can only be obtained if, in addition, the value of the function $\hat{f}'''$ is given. If one next wants to approximate locally the function $\hat{\sigma}$ and the function $\hat{\sigma}'$, it follows directly that one must perform a local third-order approximation of the function $\hat{f}$.

In this subsection we discuss three functional forms which can perform a local approximation of at least the third-order of either a given RC-function $\hat{F}$, or a given RCI-function $\hat{f}$.

It is obvious to consider first a third-order TAYLOR-approximation by means of:

$$\hat{u}(x) := a_0 + a_1 x + a_2 x^2 + a_3 x^3, \qquad (5.21)$$

or:

$$\hat{U}(X) := X_2 \hat{u}(X_1/X_2) = a_0 X_2 + a_1 X_1 + a_2 X_1^2/X_2 + a_3 X_1^3/X_2^2. \qquad (5.22)$$

The class of function $\hat{u}$, eq. (5.21), can be used to perform a third-order TAYLOR-approximation of a given RCI-function $\hat{f}$. The class of functions $\hat{U}$, eq. (5.22), can be used to perform a local third-order approximation, not

TAYLOR-approximation, of a given RC-function $\hat{F}$. The last class of functions can be considered as a direct generalization of the KUIPERS-class of functions, eq. (5.16). It is, however, a troublesome affair to investigate the restrictions on the parameters and the variable x in (5.21), such that (5.21) defines a class of RCI-functions, whose largest possible domains are bounded in $\mathbb{R}_{++}$ and for which the boundaries of each largest possible domain depend on the parameters a_0, a_1, a_2 and a_3. That is the reason that we consider the following second, directly specified, functional form:

$$\hat{U}(X) := a_1 X_1 + a_2 X_1^c X_2^{1-c} + a_3 X_3, \qquad (5.23)$$

or:

$$\hat{u}(x) := \hat{U}(x,1) = a_1 x + a_2 x^c + a_3. \qquad (5.24)$$

The class of functions defined by (5.23) has been obtained by BRUNO and VAZQUEZ, as the outcome of a model for specifying indirectly classes of production functions, see subsection 4.4.3., eq. (4.69).

It will first be proved, that this class of BRUNO-VAZQUEZ functions can be used to perform a local third-order approximation, not TAYLOR-approximation, of a given RCI-function $\hat{f}$, by considering eq. (5.24).

If it is assumed that $\hat{f}^o = \hat{u}^o$, $(\hat{f}')^o = (\hat{u}')^o$, $(\hat{f}'')^o = (\hat{u}'')^o$, $(\hat{f}''')^o = (\hat{u}''')^o$, then (5.24) can be used for the construction of the following system of equalities:

$$a_1 x^o + a_2 (x^o)^c + a_3 = \hat{f}^o, \qquad (5.25a)$$

$$a_1 + c a_2 (x^o)^{c-1} = (\hat{f}')^o, \qquad (5.25b)$$

$$c(c-1) a_2 (x^o)^{c-2} = (\hat{f}'')^o, \qquad (5.25c)$$

$$c(c-1)(c-2) a_2 (x^o)^{c-3} = (\hat{f}''')^o. \qquad (5.25d)$$

If $a_2 \neq 0$, $c \neq 0,1,2$, this system will determine a unique combination of values $\underline{c}$, $\underline{a}_1$, $\underline{a}_2$, $\underline{a}_3$ for the parameters c, a_1, a_2, a_3, namely in four steps.

If we take the ratio f''/f''', it follows from (5.25c) and (5.25d) that:

$$\underline{c} = 2 + (\hat{f}''')^o/(\hat{f}'')^o x^o \qquad (5.26)$$

The insertion of the value $c = \underline{c}$ into eq. (5.25c) yields:

$$\underline{a}_2 = (\hat{f}'')^o/(x^o)^{2-\underline{c}} \underline{c}(\underline{c} - 1). \qquad (5.27)$$

From (5.25b), (5.26) and (5.27) we obtain:

$$\underline{a}_1 = (\hat{f}')^o - \underline{c}\underline{a}_2 (x^o)^{\underline{c}-1}, \qquad (5.28)$$

and from (5.25a):

$$\underline{a}_3 = f^o - \underline{a}_1 x^o - \underline{a}_2 (x^o)^{\underline{c}}. \qquad (5.29)$$

This means that the function:

$$\hat{u}(x) := \underline{a}_1 x + \underline{a}_2 x^{\underline{c}} + \underline{a}_3, \qquad (5.30)$$

is a local second order approximation of the given RCI-function $\hat{f}$ in $x^o \in (a,b)$, see lemma 5.1.

Secondly, it can be asked for which values of the parameters a_1, a_2, a_3 and c and the variable x, eq. (5.24) defines a class of RCI-functions. That is to say, for each function $\hat{u}$, there exists an interval $[\alpha,\beta] \subset \mathbb{R}_{++}$ such that on this interval the function $\hat{u}$ is continuously twice-differentiable and $\hat{u} > 0$, $0 < x\hat{u}'/\hat{u} < 1$, $\hat{u}'' < 0$, where the interval $[\alpha,\beta]$ is the largest possible domain of the function $\hat{u}$ and the boundaries α and β depend on a_1, a_2, a_3 and c. Like in subsection 4.4.3., it can be proved that eq. (5.24) defines a class of RCI-functions, in the sense described above, if and only if the inequalities hold in one of the columns in the next table, where $\alpha := ((c-1)a_2/a_3)^{-1/c}$ and $\beta := (ca/(-a_1))^{1/(1-c)}$.

$\alpha < \beta$	$\alpha < \beta$	$\alpha < \beta$
$c < 0$	$0 < c < 1$	$c > 1$
$a_1 > 0$	$a_1 < 0$	$a_1 > 0$
$a_2 < 0$	$a_2 > 0$	$a_2 < 0$
$a_3 > 0$	$a_3 < 0$	$a_3 < 0$
$\alpha \leq x \leq \beta$	$\alpha \leq x \leq \beta$	$\alpha \leq x \leq \beta$

From this table it follows that the direct specification of the BRUNO-VAZQUEZ functional form leads to more general results than the indirect specification of the BRUNO-VAZQUEZ functional form. The indirectly specified BRUNO-VAZQUEZ functional form only generates a class of RCI-functions, with largest possible domains which are bounded in $\mathbb{R}_{++}$ and for which the boundaries of each largest possible domain depend on the parameters, if the inequalities in the third column hold.

Observe that, for $c = 2$ in eq. (5.27) one obtains the KUIPERS-class of functions, for $c = \frac{1}{2}$ the DIEWERT-class of functions (see resp. (5.16), (5.17)). From the second column it follows that the KUIPERS-class can also be specified indirectly.

As a last functional form we shall discuss a functional form which has been specified directly by DENNY (21). Let for the two-factor case the following functional form be defined by:

$$\hat{U}(X) := (a_{11}X_1^b + a_{21}X_1^{b(1-c)}X_2^{bc} + a_{12}X_1^{bc}X_2^{b(1-c)} + a_{22}X_2^b)^{1/b}, \qquad (5.31)$$

which may be written as:

$$\hat{U}(X) := \left\{ (X_1^{bc}, X_2^{bc}) \begin{pmatrix} a_{11} & a_{12} \\ a_{21} & a_{22} \end{pmatrix} \right.$$

$$\left. (X_1^{b(1-c)}, X_2^{b(1-c)}) \right\}^{1/b}. \qquad (5.32)$$

Because (5.31) can be written in the form of (5.32), DENNY called eq. (5.31) a class of Generalized Quadratic production functions (22). He noted that 'if this class of functions should only generate quasi-concave functions, it is a rather cumbersome affair to obtain a consistent set of constraints on function-variables and -parameters'. For example he showed that the constraints (i) $a_{ij} = 0$, (ii) $0 = c = 1$ and (iii) $b = 1$ are sufficient conditions for quasi-concavity of $\hat{U}$ in $\mathbb{R}^2_{++}$.

DENNY stated that (5.31) can perform a local second-order approximation of a given RC-function $\hat{F}$. Actually, DENNY's class of GQ functions can be the starting-point for a local fifth-order approximation. From the assumptions $F^o = U^o$, $(F')^o = (U')^o, \ldots, (F^{(5)}_{11111})^o = (U^{(5)}_{11111})$, one can obtain from (5.31) a system of six non-linear equations, which can be used for the computation of a unique combination of values $\underline{a}_{11}, \ldots, \underline{c}$ for the parameters of (5.31) and subsequently one can apply theorem 5.1.

From eq. (5.31) it follows that the BRUNO-VAZQUEZ-class ($b = 1$, $a_{21} = 0$) and the CES-class ($a_{12} = a_{21} = 0$) are special subclasses of DENNY's GQ class of functions. This implies that the CES-class can be the starting-point for a local second-order approximation of a given RC-function $\hat{F}$.

5.2.4. Above we reviewed several functional forms which could perform a local approximation of a given RC-function $\hat{F}$, or a given RCI-function $\hat{f}$. It must be noted that it is not possible to apply the analysis of this section to the case in which $X \notin D$.

5.5. LOCAL APPROXIMATION OF TRANSFORMS

5.5.1. In this section we shall investigate the next step in a compound approximation of a given S-neoclassical production function: local approximation of the transform W, such that the approximation V of W *is a transform*, i.e. $V(0) = 0$, V continuously differentiable function on $\mathbb{R}_{++}$ with $V' > 0$ and $\lim_{Z \to \infty} V(Z) = \infty$. Furthermore, it is our intention to connect the approximation approach of transforms with a number of functional forms which have been proposed in the literature from a different standpoint, see subsection 5.5.3. In that subsection we shall introduce the following two concepts: the function of the *scale-elasticity* and the general class of FRISCH transforms.

5.5.2. Let W be the transform of a given S-neoclassical production function H and let $Z^o = \hat{F}(X^o)$, $X^o \in D$, X^o fixed. As a first functional form which may perform a local approximation of W, we investigate the first-order TAYLOR approximation:

$$V(Z) := c_o + c_1 Z. \qquad (5.33)$$

If each function V must be a transform, then $V(0) = 0$, which implies $c_o \equiv 0$. From the assumption $V^o = W^o$ we obtain $\underline{c}_1 = W^o/Z^o$, where $\underline{c}_1$ is the value of the parameter c_1, and $V'(Z^o) = \underline{c}_1 > 0$. This means that:

1. Eq. (5.33) can only perform a first-order TAYLOR-approximation, if W is a linear function of Z, or $\underline{c}_1 = \hat{W}'(Z^o) = W^o/Z^o$, and $W(0) = 0$.
2. If eq. (5.33) is used to perform a local approximation of W and if the given RC-function $\hat{F}$ is approximated locally by means of some (linear-homogeneous) approximation $\hat{U}$ (see the preceding section), then we obtain a rather trivial compound approximation $V(\hat{U}(X))$

of the given function H: the multiplication of $\hat{U}$ by the scalar c_1. I.e., the compound function $V(\hat{U}(X))$ is homogeneous of degree one.

Since (5.33) is not very flexible, it is obvious to investigate as a next step an approximation of W by means of a second-order TAYLOR-approximation:

$$V(Z) = c_0 + c_1 Z + c_2 Z^2. \tag{5.34}$$

Again it follows from $V(0) = 0$ that $c_0 \equiv 0$. From:

$$V^o = c_1 Z^o + c_2 (Z^o)^2 = W^o, \tag{5.35a}$$
$$(V^o)' = c_1 + 2c_2 Z^o = (W')^o, \tag{5.35b}$$

we obtain a unique combination of values for the parameters c_1 and c_2 and we conclude that:

1. Eq. (5.34) performs a local first-order approximation of the transform W, see def. 5.1.
2. Actually, (5.34) defines a class of parabolas with at least one real root $Z = 0$.
3. Eq. (5.34) defines a class of transforms if and only if $c_1 \gtreqless 0$ and $c_2 > 0$, i.e. a class of strictly convex transforms.

In the same way an arbitrary n-th order TAYLOR approximation can be investigated, yielding that the n-th order polynomial has $Z = 0$ as the only non-negative root and represents a class of strictly convex transforms which can perform an $(n - 1)$-th order approximation.

Thirdly, we study the properties of the following functional form:

$$V(Z) := cZ^d, \quad c > 0, \quad d > 0, \tag{5.36}$$

a class of *power* functions with $V(0) = 0$. From:

$$V^o = c(Z^o)^d, \tag{5.37a}$$
$$(V^o)' = dc(Z^o)^{d-1}, \tag{5.37b}$$

we obtain a unique combination of values for c and d and

we conclude that:
1. This class of power functions performs a second-order approximation of a given transform W.
2. Has the properties of either a class of concave transforms ($d \leqq 1$), or a class of convex transforms ($d \geqq 1$).
3. In combination with an approximation $\hat{U}$ of a given function $\hat{F}$, eq. (5.36) defines a class of functions which is homogeneous of degree d > 0, exhibiting decreasing, or constant, or increasing returns to scale.

The three functional forms discussed above can only generate special subclasses of transforms, either convex or concave. In figures 5.1a and 5.1b we illustrate the graphs of functions which are transforms in the sense of the definition of an S-neoclassical production function, but which are not generated by eqs. (5.33), (5.34) or (5.36), (23).

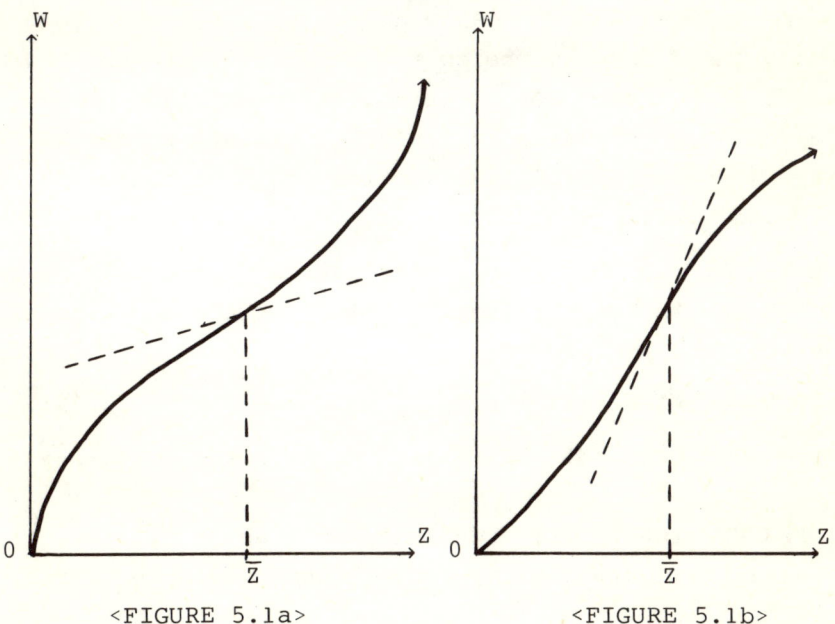

<FIGURE 5.1a>   <FIGURE 5.1b>

5.5.3. We continu our discussion of local approximation of a given transform W by introducing an additional restriction on the transform W and its approximation V. This restriction is rather complicated, but is introduced to unify views of production technologies from the past (24) with more recent (approximation-based) views. It will be assumed that the transform W and its approximation V *belongs to the class of FRISCH-transforms*.

DEFINITION 5.2: An S-neoclassical production function has a FRISCH transform, if and only if the function ε of the scale-elasticity has the following properties:
1. The function ε is positive and bounded on $\mathbb{R}_{++}$.
2. The function ε is continuously differentiable on $\mathbb{R}_{++}$ with $\varepsilon' < 0$.
3. The value of the function ε decreases monotonically from values in excess of 1, through 1 to zero.

We believe that the class of transforms W in def. 2.9, which is based on SHEPHARD's definition (25), is too general. The discussion below tries to elucidate how FRISCH starts from a subclass of transforms wich might be considered as more realistic (26). But, first, we shall discuss to concept of the function of the scale-elasticity and its implied properties.

DEFINITION 5.3: For each S-neoclassical production function H the function ε from D to $\mathbb{R}_{++}$ is the function of the *scale-elasticity* of H, if and only if:

$$\varepsilon(X) = (F_1'X_1 + F_2'X_2)/F. \qquad (5.38).$$

Associated with this definition we have the following statements:
1. FØRSUND interprets the function ε as 'a directional derivative along a ray through the origin in factor space',(27).

2. The function ε yields the value of the sum of the output elasticities, the WICKSELL/JOHNSON-theorem (28).
3. From $\varepsilon F(X) = F_1'X_1 + F_2'X_2$ it follows that the function ε might be considered as a generalization of EULER's theorem for homogeneous functions to nonhomogeneous functions. If the function H is homogeneous of a degree $\lambda > 0$, then $\varepsilon(X) \equiv \lambda$, i.e. the scale-elasticity is constant on D.

From the assumption that the function H is S-neoclassical, it follows that the function ε has a number of simple analytical properties. First, from $W' > 0$, we conclude that we can define the (unique) inverse function $T = W^{-1}$ of the transform W as follows:

$$Z = T(Y), \qquad (5.39)$$

if and only if:

$$Y = W(Z), \qquad (5.40)$$

with (i) $T(Y) > 0$ on $\mathbb{R}_{++}$ and (ii) on $\mathbb{R}_{++}$: $dT/dY = 1/W'(Z) > 0$. That is to say, there is a one-to-one correspondence between the transform W and its inverse function T. The function T will be called the *indirect transform*, because the function T has the properties of a transform. From (5.38) and (5.40) we obtain:

$$\varepsilon(X) = W'(Z)(\hat{F}_1'X_1 + \hat{F}_2'X_2)/W(Z), \qquad (5.41)$$

and from EULER's theorem and (5.39):

$$\varepsilon(X) = T(Y)dY/YdT(Y) = \varepsilon(Y). \qquad (5.42)$$

That is to say, if W is a given transform, then (i) eq. (5.39) can be used to define the indirect transform T and (ii) eq. (5.42) can be used to define the function ε of the scale-elasticity in terms of the indirect transform T, i.e. the function ε can be written as a function of the output level Y.

If in a reversed way a function $\varepsilon(Y)$ is specified, then it follows from eq. (5.42) that a class of indirect transforms $T(Y)$ can be obtained from:

$$T(Y) = \exp \int dY/Y\varepsilon(Y), \qquad (5.43)$$

which in combination with eq. (5.39) and eq. (5.40) yields a class of transforms W. It can be easily verified that if the function ε is bounded on an interval $(0,\delta)$ and is continuous on $\mathbb{R}_+$, then each function T defined by (5.43) has the properties of an indirect transform.

Eq. (5.43) will now be used for the investigation of the properties of the class of FRISCH transforms, see def. 5.2. I.e., what can be said about the properties of each transform W, if in eq. (5.43):
1. The function ε is positive and bounded on $\mathbb{R}_+$.
2. The function ε is continuously differentiable on $\mathbb{R}_{++}$ with $\varepsilon'(Y) < 0$.
3. There exists a scalar $Y^o \in \mathbb{R}_{++}$ such that:

$$\varepsilon(Y) \begin{cases} > 1 \text{ for } Y < Y^o, \\ = 1 \text{ for } Y = Y^o, \\ < 1 \text{ for } Y > Y^0. \end{cases} \qquad (5.44)$$

From an economic point of view it is important to note that it has been proved by FØRSUND that, if the function ε exhibits the properties above, then the graph of the average minimized cost function of each function W has a U-shape, if one moves along an expansion path outwards from the origin (29).

Consider an arbitrary solution T of eq. (5.43). From the properties that ε is bounded and differentiable on $\mathbb{R}_{++}$ and $\varepsilon' < 0$ we conclude that:

$$\lim_{Y \downarrow 0} T(Y) = 0, \; \lim_{Y \to \infty} T(Y) = \infty. \qquad (5.45)$$

From $\varepsilon(Y) > 0$ and (5.43) it follows that the solution T is strictly positive on $\mathbb{R}_{++}$.

From:

$$T'(Y)/T(Y) = 1/Y\varepsilon(Y) \tag{5.46}$$

we obtain that $T'(Y)$ exists and is strictly positive on $\mathbb{R}_{++}$, which implies that the direct transform $Y = W(Y)$ exists, with W strictly positive on $\mathbb{R}_{++}$, $W(0) = 0$ and $W'(Y)$ exists and is strictly positive on $\mathbb{R}_{++}$.

Differential equation (5.42) yields after differentiating and rearranging terms:

$$d\varepsilon/dY = \frac{1}{Y}(1 - \varepsilon(Y)(\frac{YT''}{T'} + 1)). \tag{5.47}$$

From the assumption that $\varepsilon'(Y) < 0$ we find:

$$YT''/T' > (1/\varepsilon(Y)) - 1. \tag{5.48}$$

For the direct transform T we obtain:

$$1 = W'(Z)/T'(Y), \tag{5.49}$$

or after differentiating with respect to Y and after rearranging terms:

$$W''(Z) = -W'(Z)T''(Y)/(T'(Y))^2. \tag{5.50}$$

From (5.44), (5.48) and (5.50) we conclude that:
1. If $Y > Y^o$, then $W''(Y) < 0$, i.e. on (Y^o, ∞) the function W is strictly concave.
2. If $Y < Y^o$, then $W''(Y)$ might be either strictly positive or strictly negative, i.e. on $(0, Y^o)$ the function is either strictly convex, or strictly concave.

This means that the function W which is illustrated in figure 5.1a is not a FRISCH transform.

As far as this we investigated in this subsection the relationship between the scale elasticity, the indirect transform and the direct transform of an S-neoclassical production function. We shall now resume the discussion on the local approximation of transforms.

It will be assumed that W is a given FRISCH transform and we shall investigate whether two different functional forms for classes of scale-elasticity functions can be the basis for the specification of classes of transforms:
1. Which may perform a local second-order approximation to the given function W.
2. Which represent classes of FRISCH transforms.

We are looking for second-order approximations of W, because the properties of the scale-elasticity function ε determine the properties of a FRISCH transform, including its first-order and second-order derivative.

ZELLNER and REVANKAR proposed the following class of scale-elasticity functions (30):

$$e(X) := 1/(a + bY). \tag{5.51}$$

It can be easily proved that (5.51) can be the basis for the specification of a class of FRISCH transforms, if $0 < a < 1$ and $b > 0$.

Eqs. (5.46) and (5.51) can be used to get the functional form of the implied class of indirect transforms S, i.e.:

$$Z = S(Y) = B^a \exp bY, \quad 0 < a < 1, \; b > 0, \tag{5.52}$$

with B a strictly positive integration constant, and:

$$\lim_{Y \to 0} S(Y) = 0, \; \lim_{Y \to \infty} S(Y) = \infty, \tag{5.53a}$$

$$S'(Y) = B(a + bY)\exp bY > 0, \tag{5.53b}$$

$$S''(Y) = B(a(a-1) + 2abY + b^2Y^2)^{a-2} \exp bY, \tag{5.53c}$$

$$S''(Y) \begin{cases} \geqq 0 \text{ for } 0 < Y \leqq (-a + \sqrt{a})/b \\ \\ < 0 \text{ for } Y > (-a + \sqrt{a})/b \end{cases} \tag{5.53d}$$

From system (5.53) we conclude that the ZELLNER-REVANKAR class of indirect transforms may be considered as regular (31), and that to this class of indirect transforms we can attribute a class of FRISCH transfroms V. It is, however, not possible to get an explicit functional form for the class of FRISCH transforms V by means of standard functional forms. That is to say, the class of functions V is defined *implicitly* by the ZELLNER-REVANKAR class of scale-elasticity functions (5.51). That is the reason that we do not investigate the approximation of the given transform W by means of the class of functions V, but we shall investigate whether it is possible to perform a local approximation of the given indirect transform $T(Y) = W^{-1}(Y)$ by means of the class of functions S (see (5.52)).

For $Y^o = W(\hat{F}(X^o))$ fixed and from:

$$S^o = B(Y^o)^a \exp bY^o = T^o, \qquad (5.54a)$$

$$(S')^o = (a + bY^o)\exp bY^o = (T')^o, \qquad (5.54b)$$

$$(S'')^o = B(a(a-1)+2abY^o+b^2(Y^o)^2)^{a-2}\exp bY^o, \quad (5.54c)$$

it is possible to obtain a unique combination of values for the parameters B, a and b of (5.52). This means that it follows form the investigations above and lemma 5.1 that eq. (5.52) can be used to perform a second-order approximation of the given function T and that, if $0 < a < 1$, $b > 0$, the class of functions S is a class of indirect transforms to which a class of FRISCH transforms can be attributed.

RINGSTAD proposed the following functional form (32):

$$e(Y) := a + b\ln Y, \quad a > 0, \quad b < 0, \quad \ln x < -a/b, \qquad (5.55)$$

and obtained:

$$S(Y) = B(a + b\ln Y)^{1/b} \qquad (5.56)$$

and:

$$V(Z) = \exp(1/b((Z/B)^b - a)). \qquad (5.57)$$

This means that:
1. Eq. (5.55) can be used to get the explicit functional forms for the functions S and T.
2. From the conditions imposed on the parameters of (5.55), the functions S and T cannot be considered as resp. an indirect transform and a FRISCH transform. Actually, eq. (5.57) has *not* the properties of a class of transforms in the sense of def. 2.9.

It can be easily proved that the implied class of RINGSTAD functions V can perform a local second-order approximation of the given transform W.

5.6. FINAL REMARKS ON LOCAL APPROXIMATION

5.6.1. In subsection 5.1.1. we argued that the direct specification of functional forms must be regarded as an essential step in empirical analyses. Furthermore we believe that the approximation approach to the direct specification of functional forms is in line with an empirical point of view: a class of functions is specified directly and one hopes that a theoretical view (the existence of an unknown functional relation between a number of economic magnitudes) can be 'approximated' by means of this class of functions. In this chapter we suggested a compound approximation approach. It must be realized that the compound functional form either may acquire a very complicated shape, or may partly not exist (e.g. the ZELLNER-REVANKAR case). This implies that either the parameters cannot be estimated by means of standard econometric methods, or must be estimated in an indirect way. There are, however, several attempts of empirical applications of homothetic production functions, exhibiting a non-constant function

of the scale-elasticity. In these applications the results of theoretical investigations of SHEPHARD are playing a very important role. These results can be summarized as follows:
1. As explained in chapter 2, under certain conditions there is a one-to-one correspondence between a production technology and a production function.
2. If all prices and the output level are given, then the minimized cost function depends on prices and the output level.
3. There exists a one-to-one correspondence between the production function and the minimized cost function, SHEPHARD's duality theorem (33).
4. The production function is homothetic, if and only if the minimized cost function is multiplicative seperable in one term the indirect transform T and the other term involving a function of input prices, (34).

NERLOVE applied the framework above in an empirical investigation concerning production functions. He started by specifying directly (35):
1. The class of functions:

$$\varepsilon(Y) := 1/(a + b\ln Y), \quad a > 0, \quad b > 0, \quad \ln Y > \frac{-a}{b} \quad (5.58)$$

for the function of the scale-elasticity, and derived:

$$T(Y) = BY^{a+b\ln(Y/2)} \quad (5.59)$$

for the first term in the minimized cost function. Observe that (5.59) is not a functional form for a class of indirect transforms to which a class of transforms W can be attributed in the sense of def. 2.9.
2. For the second term in the minimized cost function, a class of CODB-DOUGLAS functions involving input prices.

NERLOVE estimated the parameters of the class of functions T and the parameters of the class of COBB-DOUGLAS functions by means of actual data. By the application of SHEPHARD's duality theorem, he argued that the estimated values of the parameters could be used in quantifying the production function.

5.6.2. In subsection 5.3.3. we postponed the proofs of the two theorems which are constituting the mathematical basis for decompounding a local approximation of a given S-neoclassical production function H. In this subsection we present these proofs.

PROOF OF THEOREM 5.1: Let the linear operator E be defined by:

$$E := h_1(\partial/\partial X_1) + h_2(\partial/\partial X_2). \qquad (5.60)$$

Then for $p = 1, \ldots, n$, the p-th power E^p of E attributes to the given functions $u(X)$ and $v(X)$ the functions:

$$E^p(u(X)) = \sum_{k=0}^{n} \binom{p}{k}(h_1)^k(h_2)^{p-k} u_{p,k}(X), \qquad (5.61a)$$

$$E^p(v(X)) = \sum_{k=0}^{n} \binom{p}{k}(h_1)^k(h_2)^{p-k} v_{p,k}(X). \qquad (5.61b)$$

For each $X^o + h \in B_r(X^o)$ a TAYLOR-expansion of u, resp. v, can be written as follows:

$$u(X^o + h) = u^o + E(u^o) + \ldots + \frac{1}{(n-1)!}E^{n-1}(u^o) +$$
$$+ \frac{1}{n!}E^n(u(X^o + th)) \text{ for } 0 < t < 1, \qquad (5.62a)$$

$$v(X^o + h) = v^o + E(v^o) + \ldots + \frac{1}{(n-1)!}E^{n-1}(v^o) +$$
$$+ \frac{1}{n!}E^n(v(X^o + t'h)) \text{ for } 0 < t' < 1, \qquad (5.62b)$$

First it will be proved that (5.3) implies (5.2). If (5.3) holds, then from substracting (5.62a) from (5.62b) and dividing the result by $\|h\|^n$, we obtain:

$$\frac{u(X^o+h) - v(X^o+h)}{\|h\|^n} = \frac{E^n(u(X^o+th)-v(X^o+t'h))}{n! \cdot \|h\|^n} \qquad (5.63)$$

From the assumption that the partial derivatives of u and v are continuous, from $-1 \leq h_i/\|h\| \leq 1$ and from (5.63) it follows that:

$$\lim_{h \to 0}(u(X^o+h) - v(X^o+h))/\|h\|^n = 0. \qquad (5.64)$$

From (5.64) follows by definition that (5.2) holds. That (5.2) implies (5.3) will now be proven by induction. For n = 1 this implication follows directly. Assume that this implication holds for the natural number n - 1 = 1. If (5.3) holds for the number n - 1, then:

$$u(X^o+h) - v(X^o+h) = o(\|h\|^{n-1}), \qquad (5.65)$$

and from the induction-assumption:

$$u(X^o) = v(X^o), \quad u_{p,k}(X^o) = v_{p,k}(X^o), \qquad (5.66)$$

for all pairs (p,k) satisfying $1 \leq p \leq n - 1$, $0 \leq k \leq p$. From (5.66) it follows that (5.63) holds and from (5.64) that $E^p(u(X^o+th)-v(X^o+t'h))$ approaches zero if h approaches zero. Now it will be assumed that h approaches zero in a special way, namely:

$$h_1 \to 0, \quad h_2 = sh_1, \quad h_1 > 0, \qquad (5.67)$$

where $s \neq 0$, s an arbitrary scalar and $\|h\| = h_1(1+s^2)^{\frac{1}{2}}$. From system (5.61), the insertion of $h_2 = sh_1$ into (5.63), we obtain for $h_1 \to 0$:

$$\sum_{k=0}^{n} \binom{n}{k} s^{n-k}(u_{n,k}(X^o) - v_{n,k}(X^o)) = 0. \qquad (5.68)$$

The left-hand side of (5.68) is a polynomial in terms of the parameter s. Since (5.68) must hold for all s, it follows that actually (5.68) is a zero-polynomial. That is to say $u_{n,k}(X^o) = v_{n,k}(X^o)$. From this result and

(5.66) it follows that (5.2) implies (5.3). Q.E.D.

PROOF OF THEOREM 5.2: Let the functions $u^*(X)$ and $v^*(X)$ be defined by:

$$u^*(X) := f(u(X)), \quad v^*(X) := g(v(X)). \tag{5.69}$$

From these definitions we obtain for example:

$$(u^*)'_i = f'u'_i, \quad (u^*)'_i = g'v'_i, \quad i = 1,2, \tag{5.70a}$$

$$(u^*)''_{ij} = f''u'_i u'_j + f'u''_{ij}, \quad i,j = 1,2, \tag{5.70b}$$

$$(v^*)''_{ij} = g''v'_i v'_j + g'v''_{ij}, \quad i,j = 1,2. \tag{5.70c}$$

Or more generally:

$$u^*_{p,k} = f^{(p)} A_p + \ldots + f' A_1, \tag{5.71a}$$

$$v^*_{p,k} = g^{(p)} B_p + + g' B_1, \tag{5.71b}$$

for all pairs (p,k) satisfying $1 \leq p \leq 1$, $0 \leq q \leq p$, and in which A_k and B_k denote rational expressions in resp. the partial derivatives of the function u and the partial derivatives of the function v, $1 \leq k \leq p$. Since by assumption g is a local n-th order approximation of f and v is a local n-th order approximation of u, we conclude from theorem 5.1, lemma 5.1 and system (5.71) that:

$$u^*_{p,k} = v^*_{p,k}, \tag{5.72}$$

for all pairs (p,k) satisfying $1 \leq p \leq n$, $0 \leq k \leq p$. From theorem 5.1 and (5.72) we conclude that the compound function $g(v(X))$ is a local n-th order approximation of the compound function $f(u(X))$. Q.E.D.

NOTES

1. HANOCH (1975), p. 395.
2. HANOCH (1975), p. 396. HANOCH proposes the separability approach, if the number of variables is great.
3. DIEWERT (1971), (1973), (1974a), (1974b), (1976); CHRISTENSEN, JORGENSON, LAU (1973a), (1973b);

DENNY (1974); THEIL (1975), pp. 37,38; for a discussion see LAU (1974). All these investigations are in terms of functions with n variables.

4. DIEWERT (1974), p. 298; LAU (1974), p. 184 interpretes: 'the various flexible functional forms as approximations to underlying true functions rather than true functions in their own right'.
5. LAU (1974), p. 182.
6. LAU (1974), p. 182.
7. LAU (1974), p. 184.
8. BLACKORBY, RUSELL (1975), p. 31: 'Although these flexible forms can provide second order approximations, they frequently are assumed to hold exactly in order to interpret estimation results'.
9. See FØRSUND (1974), chapter V.
10. For a discussion see TAKAYAMA (1974), pp. 77, 78, 120.
11. See DIEWERT (1971) for an application to another functional form.
12. KUIPERS (1970) and KUIPERS (not dated).
13. For investigations concerning bounded substitutability see BORTS, MISHAN (1962); De BIOLLEY (1972); KNOX LOVELL (1968); KUIPERS (1973).
14. DENNY (1974), p. 24.
15. Actually DIEWERT (1971) considered a class of n-factor production functions (so-called Generalized Linear Production Functions). This class has been extended to nonhomotheticity in DIEWERT (1975). Eq. (5.17) has already been proposed by ALLEN (1964) p. 325.
16. DIEWERT (1971) presents a proof for the n-factor version.
17. For loglinear functional forms see CHRISTENSEN, JORGENSON, LAU (1973a), DIEWERT (1974b), LAU (1974).
18. LAU (1974), p. 187.
19. FUSS, McFADDEN (1971).
20. BRUNO (1968); DAGUM (1975); KADIYALA (1972).
21. DENNY (1974), with n production factors.
22. The form $x^T A y$, $x,y \in \mathbb{R}^n$, is often called a bi-linear form. This means that the DENNY class

of Generalized Bi-Linear Functions.

23. See also SHEPHARD (1970), chapter 2, figures 7 to 10.
24. FRISCH (1965). FØRSUND (1974), p.2, states: 'I have the impression that this (FRISCH) work is not as well-known as it ought to be, to judge from recent articles on this subject'.
25. SHEPHARD (1970), p. 23.
26. FRISCH (1965), rule (10.e.20), p. 170. For a discussion see FØRSUND (1974), chapters I and V.
27. FØRSUND (1974), p. 109.
28. See KRELLE (1969), p. 27.
29. See FØRSUND (1974), chapter I, section 3.2.
30. ZELLNER, REVANKAR (1969). For a survey of functional forms for scale-elasticity functions see FØRSUND (1974), p. 224. Since FØRSUND disregarded SHEPHARD's basic assumption that for each transform $W(0) = 0$, several classes of functions cannot be considered as classes of transforms.
31. Several classes of functions obtained by FØRSUND are not classes of indirect transforms and must be considered as irregular, FØRSUND (1974), chapter V.
32. RINGSTAD (1967).
33. SHEPHARD (1970), chapter 8. For discussions see DIEWERT (1974b); FØRSUND (1974), chapter 1; JORGENSON, LAU (1974a), (1974b).
34. SHEPHARD (1970), chapter 8.
35. NERLOVE (1963).

References

Allen, R.G.D. (1964), *Mathematical Analysis for Economists*, MacMillan & Co Ltd, London.
Allen, R.G.D. (1972), *Macro-Economic Theory*, MacMillan & Co Ltd, London.
Arrow, K.J., H.B. Chenery, B.S. Minhas and R.M. Solow (1961), *Capital-Labor Substitution and Economic Efficiency*, Rev. of Ec. and Stat., 43, 225-230
Bartnick, J., and B. Gahlen (1972), *CES- und VES-Produktionsfunktionen: Kritische Anmerkungen zu einer Verfehlten Kritischen Gegenüberstellung*, Kyklos, 25, 133-137.
Beckmann, M.J. (1974), *Invariant Relationships for Homothetic Production Functions*, in Eichhorn, Henn, Opitz, Shephard (eds.), 3-20.
Biolley, T. de (1972), *A New Class of Neo-Classical Production Functions with Corresponding Investment Behavior*, Rotterdam University Press, Rotterdam.
Blackorby, C., and R.B. Russell (1975), *Implicit Separability, Cost of Living Subindices, and Flexible Functional Forms*, Paper presented at the Third World Congress of the Econometric Society, Toronto.
Boer, P.M.C. de (1971), *A Note on the Relationship Between the Lu/Fletcher and the Sato/Hoffmann V.E.S. Production Functions*, Report 7114, Netherlands School of Economics, Econometric Institute, Rotterdam.
Borts, G.H., and E.J. Mishan (1962), *Exploring the "Uneconomic Region" of the Production Function*, Review of Economic Studies, 29, 300-312.
Brown, M. (1968), *On the Theory and Measurement of Technical Change*, Cambridge University Press, Cambridge.
Brown, M., and J.S. de Cani (1963), *Technical Change and the Distribution of Income*, International Economic Review, 4, 289-309.
Bruno, M. (1962), *A Note on the Implications of an Empirical Relationship Between Output per Unit of Labour, the Wage Rate, and the Capital-Labour Ratio*, Unpublished mimeo, Stanford University, Stanford.
Bruno, M. (1968), *Estimation of Factor Contribution to Growth Under Structural Disequilibrium*, International Economic Review, 9, 49-62.

Burmeister, E., and A.R. Dobell (1970), *Mathematical Theories of Economic Growth*, The MacMillan Co/Collier-MacMillan Ltd, London.

Christ, C.F. (ed.) (1963), *Measurement in Ecomomics: Studies in Mathematical Economics and Econometrics in Memory of Yehuda Grünfeld*, Stanford University Press, Stanford.

Christensen, L.R., D.W. Jorgenson and L.J. Lau (1973a), *Transcendental Logaritmic Utility Functions*, Technical Report 94, Institute for Mathematical Studies in the Social Sciences, Stanford University, Stanford.

Christensen, L.R., D.W. Jorgenson and L.J. Lau (1973), *Transcendental Logaritmic Production Functions*, Review of Economics and Statistics, 55, 28-45.

Clemhout, S. (1968), *The Class of Homothetic Isoquant Production Functions*, Review of Economic Studies, 35, 91-104.

Dagum, C. (1975), *On Constant and Variable Elasticity of Substitution Production Functions: a New Approach and an International Comparison*, Paper presented at the Third World Congress of the Econometric Society, Toronto.

Denny, M. (1974), *The Relationship Between Functional Forms for the Production System*, Canadian Journal of Economics, 7, 21-31.

Diewert, W.E. (1971), *An Application of the Shephard Duality Theorem: a Generalized Leontief Production Function*, Journal of Political Economy, 79, 481-507.

Diewert, W.E. (1973), *Functional Forms for Profit and Transformation Functions*, Journal of Economic Theory, 6, 284-316.

Diewert, W.E. (1974a), *Functional Form for Revenue and Factor Requirements Functions*, Internation Economic Review, 15, 119-130.

Diewert, W.E. (1974b), *Applications of Duality Theory*, in Intrilligator, Kendrick (eds.), 106-171.

Diewert, W.E. (1976), *Exact and Superlative Index Numbers*, Journal of Econometrics, 4, 115-145.

Eichhorn, W. (1970), *Theorie der Homogenen Productionsfunktion*, Springer-Verlag, Berlin.

Eichhorn, W., R. Henn, O. Opitz and R.W. Shephard (eds.) (1974), *Production Theory, Proceedings of an International Seminar held at the University of Karlsruhe, May-July, 1973*, Springer-Verlag, Berlin.

Eichhorn, W., and S.-C. Kolm (1974), *Technical Progress, Neutral Inventions, and Cobb-Douglas*, in Eichhorn, Henn, Opitz, Shephard (eds.), 35-45.

Ferguson, C.E. (1971), *The Neo-Classical Theory of Production & Distribution*, Cambridge University Press, London-New York.

Fleck, F.H., B. Finkbeiner and R. Casutt (1971), *Die CES- und VES-produktionsfunktion: eine Kritische Gegenüberstellung*, Kyklos, 24, 65-76.

Førsund, F.R. (1974), *Studies in the Neo-Classical Theory of Production*, Memorandum from Institute of Economics, University of Oslo, Oslo.

Frisch, R. (1965), *Theory of Production*, Reidel, Dordrecht.

Fuss, M.A., and D.L. McFadden (1971), *Flexibility versus Efficiency in Ex Ante Plant Design*, Discussion Paper 190, Harvard Institute of Economic Research, Cambridge, Mass.

Hanoch, G. (1975), *Production and Demand Models with Direct or Indirect Implicit Additivity*, Econometrica, 43, 395-419.

Hardy, G.H., J.E. Littlewood and G. Pólya (1934), *Inequalities*, Cambridge University Press, Cambridge, Mass.

Henderson, J.M., and R.E. Quandt (1958), *Microeconomic Theory*, McGraw-Hill Co. New York.

Hicks, J.R. (1932), *Theory of Wages*, MacMillan & Co, London.

Intrilligator, M.D., and D.A. Kendrick (eds.) (1974), *Frontiers in Quantative Economics*, vol. II, North-Holland Publ. Co, Amsterdam.

Jong, F.J. de, and T. Krishna Kamur (1972), *Some Considerations on a Class of Macro-Economic Production Functions*, De Economist, 120, 134-152.

Jorgenson, D.W., and L.J. Lau (1974a), *The Duality of Technology and Economic Behavior*, Review of Economic Studies, 41, 181-200.

Jorgenson, D.W., and L.J. Lau (1974b), *Duality and Differentiability in Production*, Journal of Economic Theory, 9, 32-42.

Junius, Th. (1976a), *On the Relationship Between a Homogeneous, Two-Factor, Production Function and Substitutability*, Contributed Paper presented at the European Meeting of the Econometric Society, Helsinki.

Junius, Th. (1976b), *Homothetic, Two-Factor, Production Functions: a Reconsideration of the Sato-Førsund-theorem*, Research-paper Econometric Institute, University of Groningen, Groningen.

Kadiyala, K.R. (1972), *Production Functions and Elasticity of Substitution*, Southern Economic Journal, 38, 281-284.

Knox Lovell, C.A. (1968), *Capacity Utilization and Production Function Estimation in Postwar American Manufacturing*, Quarterly Journal of Economics, 82, 219-239.

Knox Lovell, C.A. (1973), *Estimation and Prediction with CES and VES Production Functions*, International Economic Review, 14, 676-692.

Krelle, W. (1969), *Produktionstheorie*, J.C.B. Mohr (Paul Siebeck), Tübbingen.

Kuipers, S.K. (1970), *De Betekenis van Vraag- en Aanbodfactoren in Groeimodellen met één sector (The Significance of Demand and Supply Factors in One-Sector Growth Models)*, mimeographed doctoral dissertation (in Dutch), University of Groningen, Groningen.

Kuipers, S.K. (1973), *A Demand and Supply Model of Economic Growth*, De Economist, 121, 553-608.

Kuipers, S.K. (not dated), *On the Estimation of Capacity Output*, Memorandum from Institute of Economic Research, nr. 4, University of Groningen, Groningen.

Lau, L.J. (1974), *Comments on Diewert (1974b)*, in Intrilligator, Kendrick (eds.), 176-199.

Laüfer, N.K.A. (1974), *Ist die VES-Produktionsfunktion überbestimmt?*, Schweiz. Zeitschrift für Volkswirtschaft und Statistik, 2, 251-258.

Liu, T.C., and G.H. Hildebrand (1965), *Manufacturing Production Functions in the United States, 1957*, Cornell University Press, Ithaca.

Lu, Y.C. (1967), *Variable Elasticity of Substitution Production Functions, Technical Change and Factor Shares*, unpublished doctoral dissertation, Iowa State University, Ames, Iowa.

Lu, Y.C., and L.B. Fletcher (1968), *A Generalization of the CES Production Function*, Review of Economics and Statistics, 50, 449-552.

Morisett, I. (1953), *Some Recent Uses of Elasticity of Substitution - a Survey*, Econometrica, 34, 41-62.

Nerlove, M. (1963), *Returns to Scale in Electricity Supply*, in Christ (ed.), 167-198.

Paroush, J. (1964), *A Note on the CES Production Function*, Econometrica, 32, 213-214.

Paroush, J. (1966), *The h-homogeneous Production Function with Constant Elasticity of Substitution: a Note*, Econometrica, 34, 225-227.

Plasmans, J.E.J. (1975), *Production Investment Behavior*, Tilburg University Press, Tilburg.

Revankar, N.S. (1971a), *Capital-Labor Substitution, Technical Change, and Economic Growth: the U.S. Experience, 1929-1953*, Metroeconomica, 23, 154-173.

Revankar, N.S. (1971b), *A Class of Variable Elasticity of Substitution Production Functions*, Econometrica, 39, 61-71.

Ringstad, V. (1967), *Econometric Analysis Based on a Production Function with Neutrally Variable Scale-Elasticity*, Swedish Journal of Economics, 69, 115-133.

Ruys, P.H.M. (1975), *Public Goods and Decentralization*, Tilburg University Press, Tilburg.

Sato, R. (1967), *Linear Elasticity of Substitution Production Functions*, Metroeconomica, 19, 33-41.

Sato, R., and M.J. Beckmann (1968), *Neutral Inventions and Production Functions*, Review of Economic Studies, 35, 57-66.

Sato, R., and R.F. Hoffmann (1968), *Production Functions with Variable Elasticity of Factor Substitution: Some Analysis and Testing*, Review of Economics and Statistics, 50, 453-460.

Shephard, R.W. (1953), *Cost and Production Functions*, Princeton University Press, Princeton.

Shephard, R.W. (1967), *The Notion of a Production Function*, Unternehmensforschung, 11, 209, 232.

Shephard, R.W. (1970), *Theory of Cost and Production Functions*, Princeton University Press, Princeton.

Shephard, R.W. (1974), *Semi-Homogeneous Production Functions and Scaling of Production*, in Eichhorn, Henn, Opitz, Shephard (eds.), 253-285.

Shephard, R.W., and R. Färe (1974), *The Law of Diminishing Returns*, in Eichhorn, Henn, Opitz, Shephard (eds.), 278-318.

Shone, R. (1975), *Microeconomics, a Modern Treatment*, McMillan Press Ltd, London.

Solow, R.M. (1956), *A Contribution to the Theory of Economic Growth*, Quarterly Journal of Economics, 70, 65-94.

Stehling, F. (1974), *Neutral Inventions and CES Production Functions*, in Eichhorn, Henn, Opitz, Shephard (eds.), 64-94.

Takayama, A. (1974), *Mathematical Economics*, The Dryden Press, Hinsdale.

Theil, H. (1975), *Theory and Measurement of Consumer Demand*, vol. I, North-Holland Publ. Co.

Uebe, G. (1976), *Produktionstheorie*, Springer-Verlag, Berlin.

Vazquez, A. (1971), *Homogeneous Production Functions with Constant and Variable Elasticity of Substitution*, Zeitschrift für die gesamte Staatswissenschaft, 127, 7-26.

Walters, A.A. (1963), *Production and Cost Functions, an Econometric Survey*, Econometrica, 31, 1-66.

Yasui, T. (1965), *The CES Production Function: a Note*, Econometrica, 33, 646-648.

Zellner, A., and N.S. Revankar (1969), *Generalized Production Functions*, Review of Economic Studies, 36, 241-250.

Index of authors

Allen, R.G.D., 117
Arrow, K.J., 60, 62, 68, 81

Bartnick, J., 83
Beckmann, M.J., 34, 41, 44, 49
Biolley, T. de, 83, 117
Blackorby, C., 117
Boer, P.M.C. de, 83
Borts, G.H., 33, 117
Brown, M., 60, 61, 81, 82
Bruno, M., 71, 72, 76, 81, 99, 102
Burmeister, E., 35, 36, 56, 57

Cani, J.S., de, 60, 61, 81
Casutt, R., 83
Chenery, H.B., 60, 62, 68, 81
Christensen, L.R., 116, 117
Clemhout, S., 83

Dagum, S., 117
Denny, M., 82, 101, 117
Diewert, W.E., 8, 82, 95, 101, 116, 117, 118
Dobell, R.E., 35, 36, 56, 57

Eichhorn, W., 33, 56, 82

Färe, R, 33
Ferguson, C.E., 32, 33, 83
Finkbeiner, B., 83
Fleck, F.H., 83
Fletcher, L.B., 83
Førsund, F.R., 25, 32, 57, 106, 108, 118
Frisch, R., 32, 106, 118
Fuss, M.A., 97

Gahlen, B., 83
Hanoch, G., 84, 85
Hardy, G., 82
Henderson, J.M., 32
Henn, R., 82
Hicks, J.R., 32
Hildebrand, G.H., 72, 81
Hoffmann, R.F., 74, 80, 81, 83

Jong, F.J. de, 82
Jorgenson, D.W., 116, 117, 118
Junius, Th., 31, 33

Kadiyala, K.R., 67, 117
Knox Lovell, C.A., 83, 117
Kolm, S.-C., 56
Krelle, W., 32, 50, 118
Krishna Kamur, T., 82
Kuipers, S.K., 2, 33, 95, 118

Lau, L.J., 87, 88, 97, 116, 117, 118
Laüder, N.K.A., 83
Littlewood, J.E., 82
Liu, T.C., 72, 81
Lu, Y.C., 82

McFadden, D.L., 97
Minhas, B.S., 60, 62, 68, 81
Mishan, E.J., 33, 117
Morisett, I., 32

Nerlove, M., 113

Opitz, O., 82

Paroush, J., 81, 82
Plasmans, J.E.J., 2, 16, 32
Pólya, G., 82

Quandt, R.E., 32

Revankar, N.S., 66, 81, 110
Ringstad, V., 111
Rusell, R.B., 117
Ruys, P.H.M., 10, 31

Sato, R., 25, 56, 60, 66, 80, 81, 82
Shephard, R.W., 2, 9, 31, 32, 33, 34, 82, 113, 118
Shone, R., 8, 31, 32
Solow, R.M., 38, 60, 62, 68, 81
Stehling, F., 56, 57

Takayama, A., 1, 8, 117
Theil, H., 117

Uebe, G., 32

Vazquez, A., 32, 76, 81, 82

Walters, A.A., 8

Yasui, T., 82

Zellner, A., 110

Index of subjects

activity-analysis, 1, 10
approximation, 86
  decomposition of
    local -, 88
  global -, 87
  local, 87, 88
  local - of RC- and
    RCI-functions, 92
  local - of trans-
    forms, 103
  Taylor -, 92, 98, 103
asymmetry, 95
average minimized costs,
  108, 113
  multiplicative sepa-
    rability of -, 113
  U-shape of -, 108

bounded substitutability,
  95, 117

capital-labor ratio, 36
characteristic equation,
  54
competitive equilibrium
  - of factor markets,
    36, 63, 73, 81
  - of output market,
    73, 81
compound function, 4, 17,
  90, 92
cone, 16
core function, 15, 16
  restricted -, 20
core-intensity function
  restricted -, 21

disequilibrium factor
  market, 71, 78
duality theorem, 113

efficient subset, 10
  bounded -, 11
  largest possible -, 37,
    58, 85
elasticity
  dynamic - of substitu-
    tion, 44
  - of substitution, 24
  output -, 70, 107
  scale -, 106
empirical relation, 63,
  64, 81
equivalence-theorem, 15
Euler's theorem, 21, 73,
  107

factor shares
  constant marginal -, 71
  relative -, 49

growth rate, 38
  balanced -, 41

Harrod-Domar consumption
  function, 38

implicit function, 45, 51,
  111
income share, 75
input ratio, 21

kernel function, 15

law of diminishing returns
  to factors, 30
level set, 14

marginal rate of substitution, 23
  dynamic -, 43
multi function, 10

neoclassical one-sector
  growth-model, 35
  fundamental equation
  of -, 38
  equilibrium value
  of -, 39

power function, 104
production isoquant, 13
production factors
  essential -, 18
  nonessential -, 62
production function, 14
  CES -, 12, 35, 59, 102
  Cobb-Douglas -, 42
  dynamic -, 42
  homogeneous -, 32, 62, 105
  homothetic -, 15, 113
  linear -, 62
  neoclassical -, 16
  restricted -, 20
  S-neoclassical, 19
  VES -, 65
  Walras-Leontief -, 62
production input set, 10
production technology, 11
  Shephard -, 11
  Walras-Leontief -, 13

RC-function, 20
RCI-function, 21
RP-function, 20

separability, 113, 116
specification
  direct -, 84
  indirect -, 60
stable
  globally -, 39
  locally asymptoti-
  cally -, 39

technical change, 36, 42
  generalized Beckmann
  neutral -, 49
  generalized Hicks
  neutral -, 44
  output augmenting -, 43
  output and capital
  augmenting -, 43

time-dependence, 52
transform, 15
  Frisch -, 106

Wicksell-Johnson theorem, 107

AUG 21 '84